新编高等职业教育电子信息、机电类规划教材·机电一体化技术专业

机械专业交际英语（第2版）

Communicative English for Mechanical Purposes（The second edition）

汤彩萍　　　编著

Satina Anziano（美）　主审

電子工業出版社

Publishing House of Electronics Industry

北京·BEIJING

内 容 提 要

本书基于机械类专业工作岗位对英语能力的需求，立足“能力本位”，以交际性为特色，突出专业英语的听说口语表达能力和专业应用文写作能力的训练，并以此为推手促进学习者专业英语阅读与翻译能力及学习兴趣的提高。

本书共分 8 个学习情境，是根据机械行业的典型工作情境和职业场景设计的。学习情境从机械工人职业就业展开，学习者以普通车床和数控加工中心为主要学习设备来学习机床的相关操作，追随当今机械制造业的高度自动化和网络化趋势，领略自动化工厂的运行，参加国际机床展览会，开展技术交流与合作，最后以毕业生就业面试而结束课程的学习。全书前后呼应，融会贯通；情境的听、说、读、写、译各部分紧扣同一主题展开，任务驱动，行动导向。

本书配备光盘，包含阅读和视、听、说等多媒体教学材料，以此提高教学效率。

本书适合职业院校、成人高校的机械制造与自动化类专业、数控技术类专业、机电设备类专业选用，也适合从事机械技术或产品营销的社会人员学习使用，是训练机械类专业英语听说能力的好帮手。

图书在版编目（CIP）数据

机械专业交际英语 / 汤彩萍编著. —2 版. —北京：电子工业出版社，2016.8

ISBN 978-7-121-29674-1

Ⅰ. ①机… Ⅱ. ①汤… Ⅲ. ①机械工程－英语－高等学校－教材 Ⅳ. ①TH-43

中国版本图书馆 CIP 数据核字（2016）第 189758 号

策　　划：陈晓明
责任编辑：郭乃明　　特约编辑：范　丽
印　　刷：北京捷迅佳彩印刷有限公司
装　　订：北京捷迅佳彩印刷有限公司
出版发行：电子工业出版社
　　　　　北京市海淀区万寿路 173 信箱　邮编：100036
开　　本：787×1092　1/16　印张：16.25　字数：416 千字
版　　次：2011 年 2 月第 1 版
　　　　　2016 年 8 月第 2 版
印　　次：2022 年 1 月第 7 次印刷
定　　价：39.00 元（含光盘 1 张）

凡所购买电子工业出版社图书有缺损问题，请向购买书店调换。若书店售缺，请与本社发行部联系，联系及邮购电话：（010）88254888，88258888。

质量投诉请发邮件至 zlts@phei.com.cn，盗版侵权举报请发邮件至 dbqq@phei.com.cn。

本书咨询联系方式：010-88254561。

前　言

高职英语教育是培养技能型人才的高等英语教育，旨在训练其在生产、管理、服务第一线所需的语言交际能力和应对各种涉外局面的语言应用能力。在当今职业教育大发展大改革的浪潮下，高职英语课程应根据本专业工作岗位的实际需要，立足“能力本位”，以综合职业能力培养为目的，工学结合，帮助学生掌握本专业听、说、读、写、译的语言基本技能，实现高职人才的培养目标。

本教材顺应潮流，颠覆了原有专业英语只强调阅读理解和翻译能力培养的教学理念，而以交际应用为目的设计学习情境，突出专业英语的听说口语表达能力和专业应用文写作能力的训练，并以此为推手促进学生阅读翻译能力和学习兴趣的提高。

1．情境设计

本教材根据机械行业高职人才的典型工作情境和职业场景设计了 8 个学习情境（Learning Situation，LS）。学习情境从机械工职业就业展开，学习者以普通车床和数控加工中心为主要学习设备，学习机床的操作，追随当今机械制造业的高度自动化和网络化的趋势，领略自动化工厂的运行，参加国际机床展览会，开展技术交流与合作，最后以毕业生就业面试而结束课程的学习。全书前后呼应，融会贯通。

2．情境架构

各情境编写框架示意图如下：

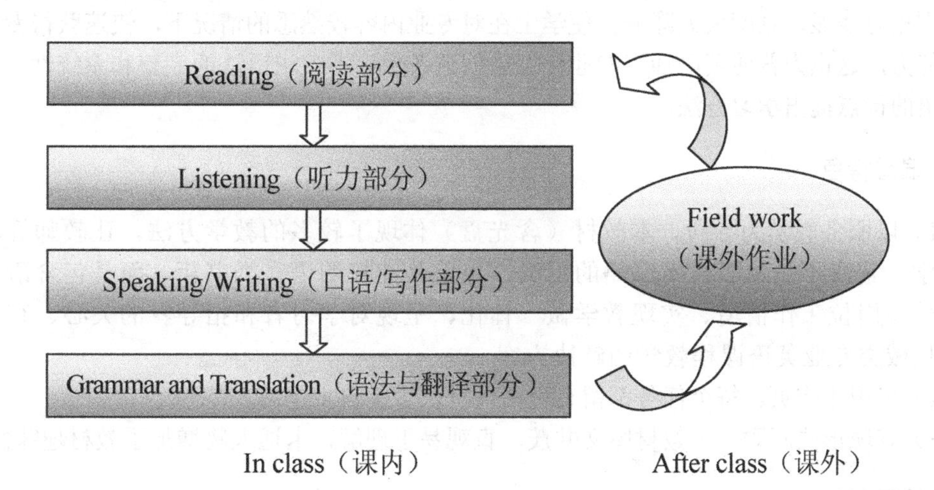

情境的各部分紧扣同一主题展开，Reading 为后续 Listening 和 Speaking 提供专业背景材料和专业词汇储备，Field work 的完成时间贯穿整个情境的学习过程。

3．教学理念和方法

本教材体现以下教学理念和教学方法：

（1）听说交际。遵循语言教学规律，以岗位需要的听说能力训练为抓手展开英语教学。听力材料的选取围绕情境主题，选择原汁原味的英文原声录像。口语任务的设计呈现递进层次，如看图说话→小对话→情境对话。

（2）任务驱动，行动导向。借鉴普通英语教学法设计了丰富的听、说、读、写、译学习任务或趣味游戏活动（如小组竞赛、角色表演、配音比赛等），以专业内容为载体训练学生的英语综合应用能力，使学生成为学习的主体。任务来自于岗位工作分析，如实际技术交流中经常会遇到符号或数字的读法问题，因此在 LS3 中设计了数字阅读的口语题。

任务形式多样：阅读训练题型达二十种，听力训练主要是根据视听材料完成填空、选择、回答问题等任务，口语训练主要有看图说话、互相提问和回答、情境对话、PPT 汇报、视频配音等，写作训练有英文摘要和英文简历书写，翻译有专业材料的英译中或中译英。

（3）多媒体教学。本教材配备光盘，其中包含了课文朗读、生词发音、视听和口语材料，是实现多媒体教学的支撑。

（4）综合职业能力培养。以结对或小组开展学习，培养合作能力、表达能力。Reading 部分注重专业能力的培养，而 Listening 和 Speaking 部分则侧重方法能力和社会能力的培养。

（5）语言应用性。本教材将普通英语与专业有机融合，学习者既学语言，又学专业，用英语表达专业（听说写），用英语寻求技术答案（读译）。

（6）边学边做边说。引入企业员工培训理念，借鉴日本产业训练协会 TWI 工作指导方法（Job Instruction）训练学生口语，如根据课文说明操作机床，边做边说。

（7）遵循外语教学规律。由于外语学习有较多的记忆任务，采用机械式背诵记忆令学生感到乏味；本教材对于同一学习要点，采用了形式多样的任务形式，使学生在反复训练中掌握了学习重点。

（8）学习难度。本教材与传统的专业英语教材相比，突出了英语的交际特色，而专业性内容则相对浅显，难度大大降低；使学生在对专业内容较熟悉的情况下，快速获得专业英语应用能力，这也为普通英语师资专业化发展创造条件。不追求语法的完整和系统性，只就专业英语的特点提出学习方法。

4．编写特色

（1）以服务教学为导向。本教材（含光盘）体现了较多的教学方法，让教师好教，学生易学。它既不是原版技术书籍的汇编，也不是纯粹的语言类书籍，而是包含语言和专业、体现岗位工作情境、实现教学做一体化、呈现对学习者和指导者的关心、有效实施高职机械类专业英语课程教学的解决方案。

（2）采用工作页。每个任务都留有空白页，供学生自主学习。

（3）表现形式新颖。本教材图文并茂，直观易于理解，卡通人物增加了教材趣味性。

5．适用范围

本教材适合职业院校、成人高校的机械制造与自动化类专业、数控技术类专业、机电设备类专业选用，也适合从事机械技术或产品营销的社会人员学习使用，尤其是训练机械类专业口语和听说能力的好帮手。本教材建议总学时为 64，可根据实际专业课程设置情况选学部分内容。

6. 致谢

本教材由美籍语言学硕士 Satina Anziano 女士担任主审。在编写过程中得到了常州创胜特尔数控机床设备有限公司赖立迅总工程师、德国博世力士乐公司 Tim Hohlmann 先生和吴宏娟女士的大力帮助，并得到了这些单位的支持；还得到了常州机电职业技术学院许朝山、苌晓兵、于华艳、靳敏、金志国、陶波等老师的热情帮助，在此一并表示感谢。

由于作者学识和经验有限，书中难免有错误与不妥之处，恳请使用者批评指正！同时，作者为用户提供了电子教案、教学指南、参考答案、视听文字稿等丰富的配套资源，扫描封面二维码即可获得。作者联系方式：13861274230（电话）。

汤彩萍
2016 年 1 月

Contents

Learning Situation 1

What machinists do

Focus of the situation

This class will discuss the job employment of machinists in the mechanical manufacturing field in terms of work procedures, job safety, and job prospects. [本课从工作内容、工作安全、就业前景等方面讨论机械制造领域机械工的工作与就业。]

Field work

How much do you know about employment opportunities related to the mechanical trade? What are you going to do in the future? Make a PPT and prepare for an in-class presentation.

Part A Reading

Machinists use machine tools, such as lathes, milling machines, and machining centers, to produce precision metal parts[1]. Although they may produce large quantities of one part, precision machinists often produce small batches or one-of-a-kind items. They use their knowledge of the working properties of metals and their skill with machine tools to plan and carry out the operations needed to make machined products that meet precise specifications[2].

Before they machine a part, machinists must carefully plan and prepare the operation[3]. These workers first review blueprints or written specifications for a job. Next, they calculate where to cut or bore into the workpiece, how fast to feed the metal into the machine, and how much metal to remove. They then select tools and materials for the job, plan the sequence of cutting and finishing operations, and mark the metal stock to show where cuts should be made[4].

After this layout work is completed, machinists perform the necessary machining operations. They position the metal stock on the machine tool—drill press, lathe, milling machine, or other type of machine—set the controls, and make the cuts. During the machining process, they must constantly monitor the feed rate and speed of the machine. Machinists also ensure that the workpiece is being properly lubricated and cooled, because the machining of metal products generates a significant amount of heat. The temperature of the workpiece is a key concern because most metals expand when heated; machinists must adjust the size of their cuts relative to the temperature[5]. Some rare but increasingly popular metals, such as titanium, are machined at extremely high temperatures.

Machinists detect some problems by listening for specific sounds—for example, a dull cutting tool or excessive vibration. Dull cutting tools are removed and replaced. Cutting speeds are adjusted to compensate for harmonic vibrations, which can decrease the accuracy of cuts, particularly on newer high-speed spindles and lathes[6]. After the work is completed, machinists use both simple and highly sophisticated measuring tools to check the accuracy of their work against blueprints.

CNC machinists

Some machinists, often called production machinists, may produce large quantities of one part, especially parts requiring the use of complex operations and great precision. Many modern machine tools are computer numerically controlled (CNC). Frequently, machinists work with computer-control programmers to determine how the automated equipment will cut a part. The programmer may determine the path of the cut, while the machinist determines the type of the cutting tool, the speed of the cutting tool, and the feed rate. Because most machinists train in CNC programming, they may write basic programs themselves and often modify programs in response to problems encountered during test runs[7]. After the production process is designed, relatively simple and repetitive operations normally are performed by machine setters, operators, and tenders.

Some manufacturing techniques employ automated parts loaders, automatic tool changers, and computer controls, allowing machine tools to operate without anyone present. One production machinist, working 8 hours a day, might monitor equipment, replace worn cutting tools, check the accuracy of parts being produced, and perform other tasks on several CNC machines that operate

24 hours a day (lights-out manufacturing). During lights-out manufacturing, a factory may need only a few machinists to monitor the entire factory.

Maintenance machinists

Other machinists, often called mechanics, do maintenance work—repairing or making new parts for existing machinery. To repair a broken part, maintenance machinists may refer to blueprints and perform the same machining operations needed to create the original part.

Work safety

Today, most machine shops are relatively clean, well lit, and ventilated. Many computer-controlled machines are partially or totally enclosed, minimizing the exposure of workers to noise, debris, and the lubricants used to cool workpieces during machining. Nevertheless, working around machine tools presents certain dangers, and workers must follow safety precautions. Machinists wear protective equipment, such as safety glasses to shield against bits of flying metal and earplugs to dampen machinery noise. They also must exercise caution when handling hazardous coolants and lubricants, although many common water-based lubricants present little hazard. The job requires stamina, because machinists stand most of the day and, at times, may need to lift moderately heavy workpieces.

Job opportunities

Job opportunities for machinists should continue to be good, as employers value the wide-ranging skills of these workers. Also, many young people prefer to attend college or may not wish to enter production occupations. Therefore, the number of workers learning to be machinists is expected to be less than the number of job openings arising each year from the need to replace experienced machinists who retire or transfer to other occupations[8].

So don't hesitate any longer, and let's get into the machine shop right now.

TECHNICAL WORDS

machinist	[məˈʃiːnist]	*n.*	机械工，机械师
lathe	[leið]	*n.*	车床
mill	[mil]	*v. & n.*	铣削；铣刀；铣床；工厂
machine	[məˈʃiːn]	*n. & v.*	机器，机械；机床；机加工
precision	[priˈsiʒən]	*n.*	精密，精度
property	[ˈprɔpəti]	*n.*	属性，特性
specification	[ˌspesifiˈkeiʃən]	*n.*	规格

blueprint	[ˈbluːˌprint]	*n.*	蓝图
job	[dʒɔb]	*n.*	工作（任务），作业，零件活
bore	[bɔː]	*v.*	镗孔，钻孔
workpiece	[ˈwəːkpiːs]	*n.*	工件，加工件
feed	[fiːd]	*n. & v.*	进给，切入
stock	[stɔk]	*n.*	毛坯，余量；库存
position	[pəˈziʃən]	*n. & v.*	位置，岗位；定位
drill	[dril]	*v. & n.*	钻削；钻头；钻床
lubricate	[ˈluːbrikeit]	*v.*	润滑
cool	[kuːl]	*v.*	冷却
vibration	[vaiˈbreiʃən]	*n.*	振动
compensate	[ˈkɔmpənseit]	*v.*	补偿
accuracy	[ˈækjurəsi]	*n.*	精度
spindle	[ˈspindl]	*n.*	主轴
modify	[ˈmɔdifai]	*v.*	修改
process	[ˈprəuses]	*n.*	工艺流程
manufacture	[ˌmænjuˈfæktʃə]	*v.*	制造
technique	[tekˈniːk]	*n.*	技术，技巧，方法
employ	[imˈplɔi]	*v.*	雇用，使用
operator	[ˈɔpəreitə]	*n.*	操作员
maintenance	[ˈmeintinəns]	*n.*	维护，保养；维修
machinery	[məˈʃiːnəri]	*n.*	机械，机器（不可数名词）
create	[kriˈeit]	*v.*	制造，创建
mechanic	[miˈkænik]	*n.*	机修工
ventilate	[ˈventileit]	*v.*	通风
lubricant	[ˈluːbrikənt]	*n.*	润滑液（剂）
coolant	[ˈkuːlənt]	*n.*	冷却液

PHRASES

machine tool	机床
milling machine	铣床
machining center	加工中心
small batches or one-of-a-kind items	单件小批量
finishing operation	精加工操作
drill press	台式钻床

feed rate		进给速度
cutting tool		刀具
cutting speed		切削速度
harmonic vibration		谐振
computer numerically controlled	(CNC)	数控
computer-control programmer		数控编程员
test run		试运行
automatic tool changer	(ATC)	自动换刀装置
computer control		计算机数控系统
machine setter		机床调试工
machine tender		机床放料工
worn cutting tool		磨损刀具
lights-out manufacturing		无人值守制造，自动化制造
machine shop		车间
computer-controlled machine		数控机床
safety precaution		安全预防
safety glasses		安全眼镜

NOTES

1. Machinists use **machine tools**, such as lathes, milling machines, and machining centers, to produce precision metal parts. 机械工使用诸如车床、铣床和加工中心等机床制造精密金属零件。*机床是制造各种机器（machine）的工具，被称为工业母机，因此英语里的"机床"往往用"machine tool"来表达。*

2. They use their knowledge of the working properties of metals and their skill with machine tools to plan and carry out the operations **needed** to make **machined** products **that meet precise specifications**. 他们运用其金属材料特性方面的知识和机床方面的技能进行工艺规划和加工，制造满足精度要求的机加工产品。*needed 为过去分词，作为后置定语，修饰operations，相当于定语从句that are needed；that meet precise specifications 是定语从句，修饰products。machined products 中machined 是过去分词作为定语。*

3. Before they machine a part, machinists must carefully plan and prepare the operation. 加工零件之前，机械工必须对整个加工过程进行仔细的规划和准备。

4. These workers **first** review blueprints or written specifications for a job. **Next**, they calculate where to cut or bore into the workpiece, how fast to feed the metal into the machine, and how much metal to remove. They **then** select tools and materials for the job, plan the sequence of cutting and finishing operations, and mark the metal stock to show where cuts should be made. 首先，这些工人阅读作业零件的图纸或书面说明；接下来，他们计算切入或钻入工件的位置、工件的进给速度、金属的去除量；然后，他们选择适合工件的刀具和材料，制定粗加工和精

加工操作的顺序，在金属毛坯上画线标记切削位置。*需要表达工作步骤时，可以用 first…next…then…或 before… after…*

5. The temperature of the workpiece is a key concern because most metals expand **when heated**; machinists must adjust the size of their cuts relative to the temperature. 因为大多数金属受热后都会膨胀，所以工件的温度是主要关注的问题；机械工必须基于温度调整切削用量。*when heated 相当于 when they are heated，heated 为过去分词，表示被动。*

6. Cutting speeds are adjusted to compensate for harmonic vibrations, **which can decrease the accuracy of cuts**, particularly on newer high-speed spindles and lathes. 调整切削速度以补偿谐振造成的误差，谐振会降低切削精度，尤其在一些新型的高速主轴和车床上。*which can decrease the accuracy of cuts 是定语从句，修饰 harmonic vibrations。*

7. Because most machinists train in CNC programming, they may write basic programs themselves and often modify programs in response to problems **encountered** during test runs. 由于大多数机械工都接受过数控编程的培训，他们会编写基本的程序。在试运行过程中，他们经常修改程序以应对碰到的问题。*encountered 为过去分词，作为后置定语，修饰 problems。*

8. Therefore, the number of workers **learning to be machinists** is expected to be less than the number of job openings **arising each year from the need to replace experienced machinists who retire or transfer to other occupations.** 每年都有一些有经验的机械师退休或跳槽到其他职业，因此带来的工作空缺数量大于准备从事机械工工作的工人数量。*learning to be machinists 是 workers 的定语，arising each year from the need 是 job openings 的定语，to replace experienced machinists 是 need 的定语，定语从句 who retire or transfer to other occupations 修饰 experienced machinists。这句话结构比较复杂，翻译比较困难，但只要通过分析语法，就能理解其含义，然后再意译成通顺的汉语。*

PRACTICE

Task 1　Translate the following words or phrases into English.

1．精密零件	2．加工零件(*v.*)
3．精加工	4．金属毛坯
5．进给速度	6．更换刀具
7．维修机械工	8．符合规格
9．修改程序	10．遇到问题
11．数控机床	12．加工中心
13．车床	14．铣床
15．镗床	16．钻床
17．强烈的振动	18．完成功能
19．完成操作	20．完成任务
21．冷却(*n.*)	22．润滑(*n.*)
23．高精度	24．高速主轴
25．切削路线	26．全封闭

27．无人化制造　　　　28．数控系统

29．数控编程（员）　　　　30．数控操作（员）

Task 2　Choose the correct English explanation for each of the following words.

1. Lathe

(A) is a machine that turns a piece of metal round and round against a sharp tool that gives it shape.

(B) is the motion of moving the work piece and the cutting tool together so as to remove material.

(C) is the operation of enlarging a hole with a single-point tool. This operation produces a close tolerance（公差） and fine（精细的） finish（表面光洁度）.

Your answer: __________

2. Milling machine

(A) is a machine that turns a piece of metal round and round against a sharp tool that gives it shape.

(B) is a machine that removes metal through the use of electrical sparks（电火花）which burn away the metal.

(C) is a machine tool that removes material by rotating a cutter and moving into the material. It is used to produce flat and angular surfaces, grooves（槽）, contours（轮廓）, and gears.

Your answer: __________

3. Boring

(A) is a machine that turns a piece of metal round and round against a sharp tool that gives it shape.

(B) is the process or technique of reducing wear（磨损）between surfaces by using a lubricant between the surfaces.

(C) is the operation of enlarging a hole with a single-point tool. This operation produces a close tolerance and fine finish.

Your answer: __________

4. Feed

(A) is the motion of moving the work piece and the cutting tool together so as to remove material.

(B) is the process or technique of reducing wear between surfaces by using a lubricant between the surfaces.

(C) is the operation of enlarging a hole with a single-point tool. This operation produces a close tolerance and fine finish.

Your answer: __________

5. CNC

(A) is a special liquid that performs three main functions during machining. It lubricates the cutting action, carries off the heat generated, and flushes（冲洗）the chips（切屑）.

(B) is a form of programmable automation in which the machine tool is controlled by a program in computer memory.

(C) is the process of removing metal with machine tools such as lathes, mills and a wide variety of other tools.

Your answer: ___________

6. Lubrication

(A) is the process or technique of reducing wear between surfaces by using a lubricant between the surfaces.

(B) is a special liquid that performs three main functions during machining. It lubricates the cutting action, carries off the heat generated, and flushes the chips.

(C) is the material being machined. It can be any material and any shape. In the machine shop it usually refers to round or flat pieces of metal ready to be machined.

Your answer:___________

7. Coolant

(A) is the process or technique employed to reduce wear between surfaces by using a lubricant between the surfaces.

(B) is a special liquid that performs three main functions during machining. It lubricates the cutting action, carries off the heat generated, and flushes the chips.

(C) is the material being machined. It can be any material and any shape. In the machine shop it usually refers to round or flat pieces of metal ready to be machined.

Your answer:___________

8. Stock

(A) is the process or technique employed to reduce wear between surfaces by using a lubricant between the surfaces.

(B) is that portion（一部分） of a machine tool that spins（旋转）either the workpiece or the cutting tool and is driven by the motor. On a milling machine, it turns within the quill（套筒）while on a lathe it turns within the headstock（床头箱）.

(C) is the material being machined. It can be any material and any shape. In the machine shop it usually refers to round or flat pieces of metal ready to be machined.

Your answer:___________

9. Spindle

(A) is a special liquid that performs three main functions during machining. It lubricates the cutting action, cools the cutting action, and flushes the chips.

(B) is that portion of a machine tool that spins either the workpiece or the cutting tool and is driven by the motor. On a milling machine it turns within the quill while on a lathe it turns within the headstock.

(C) is the material being machined. It can be any material and any shape. In the machine shop it usually refers to round or flat pieces of metal ready to be machined.

Your answer:

10. Machining

(A) is a person who uses machine tools to make or modify parts, primarily metal parts.

(B) is the process of removing metal with machine tools such as lathes, mills and a wide variety of other tools.

(C) is the piece of metal that is being shaped.

Your answer:___________

Task 3 Fill in the brackets with words that have similar meaning to the underlined words, changing their forms if necessary.

1. (   ) Machinists use machine tools, such as lathes, milling machines, and machining centers, to produce precision metal parts.

2. (   ) Machinists use machine tools, such as lathes, milling machines, and machining centers, to produce precision metal parts.

3. (   ) They use their knowledge of the working properties of metals and their skill with machine tools to plan and carry out the operations needed to make machined products that meet precise specifications.

4. (   ) One production machinist might monitor equipment, replace worn cutting tools, check the accuracy of parts being produced, and perform other tasks on several CNC machines.

5. (   ) After the work is completed, machinists use both simple and highly sophisticated measuring tools to check the accuracy of their work against blueprints.

6. (   ) Machinists may write basic programs themselves and often modify programs in response to problems encountered during test runs.

7. (   ) Machinists may write basic programs themselves and often modify programs in response to problems encountered during test runs.

8. (   ) Machinists also ensure that the workpiece is being properly lubricated and cooled, because the machining of metal products generates a significant amount of heat.

9. (   ) Some manufacturing techniques employ automated parts loaders, automatic tool changers, and computer controls, allowing machine tools to operate without anyone present.

10. (   ) Other machinists do maintenance work—repairing or making new parts for existing machinery.

11. (   ) The job requires stamina（体力）, because machinists stand most of the day and, at times, may need to lift moderately heavy workpieces.

12. (   ) Machinists must exercise caution when handling hazardous coolants and lubricants, although many common water-based lubricants present little hazard.

13. (   ) Machinists must exercise caution when handling hazardous coolants and lubricants, although many common water-based lubricants present little hazard.

14. (   ) To repair a broken part, maintenance machinists may refer to blueprints and perform the same machining operations needed to create the original part.

15. (   ) During lights-out manufacturing, a factory may need only a few machinists to monitor the entire factory.

16. (   ) They design and carry out the operations needed to make machined products that meet precise specifications.

17. (   ) Machinists use machine tools, such as lathes, milling machines, and machining centers, to produce precision metal parts.

18. (   ) Machinists use both simple and highly sophisticated measuring tools to check the accuracy of their <u>work</u> against blueprints.

19. (   ) The temperature of the workpiece is a key concern because most metals expand when heated; machinists must adjust the size of their cuts <u>relative to</u> the temperature.

20. (   ) These workers first review <u>blueprints</u> or written specifications for a job.

Task 4 The following information relates to what machinists do. Match Column A with Column B.

Column A	Column B
use	feeds and speeds
produce	machine tools
review	the accuracy
calculate	the metal stock
position	blueprints
set	problems
monitor	dull tools
detect	the machining
replace	the controls
check	parts

Task 5 Fill in the blanks with the following words, changing their forms if necessary.

machine setter, CNC, workpiece, lathe, plan, check, perform, accuracy

1. ____________ is the piece of metal that is being shaped.

2. ____________ refers to a computer "controller" that reads code instructions and drives the machine tool.

3. The working procedures for machinists are: first, they must carefully _______ and prepare the operation; next, they _______ the necessary machining operations; then they _______ the accuracy of their work against blueprints.

4. ____________ is a machine that turns a piece of metal round and round against a sharp tool that gives it shape.

5. ____________ prepare the machines prior to production, perform initial test runs producing a part, and may adjust and make minor repairs to the machinery during its operation.

6. After the work is completed, machinists use both simple and highly sophisticated measuring tools to check the ____________ of their work against blueprints.

Task 6 Choose the best answer.

1. Before they machine a part, the first thing for machinists is:

(A) To review blueprints or written specifications for a job

(B) To calculate where to cut or bore into the workpiece, how fast to feed the metal into the machine, and how much metal to remove

(C) To select tools and materials for the job, plan the sequence of cutting and finishing operations, and mark the metal stock to show where cuts should be made

Your answer:___________

2. The proper sequence of performing the necessary machining operations for machinists is as the following:

① complete layout work
② set the controls
③ position the metal stock on the machine tool
④ monitor the feed rate and speed of the machine
⑤ make the cuts

(A) ①→③→②→⑤→④

(B) ①→②→④→③→⑤

(C) ①→②→③→④→⑤

Your answer: __________

3. Cutting speeds are reduced to compensate for harmonic vibrations, which can_____the accuracy of cuts.

(A) Increase

(B) Decrease

(C) Neither of the above

Your answer: __________

4. Sounds in the machine tool are probably caused by:

(A) A dull cutting tool

(B) High speeds

(C) A dull cutting tool or high speeds

Your answer: __________

5. Who determines the path of the cut?

(A) The machinist.

(B) The programmer.

(C) The operator.

(D) The machine setter.

Your answer: __________

6. Who determines the type of the cutting tool, the speed of the cutting tool, and the feed rate?

(A) The machinist.

(B) The programmer.

(C) The machine tender.

(D) The machine setter.

Your answer: __________

7. Whom are relatively simple and repetitive operations normally performed by?

(A) The machinist. (B) The programmer.

(C) The machine operator. (D) The machine maintenance person.

Your answer: ________

8. During lights-out manufacturing,_______.

(A) Works operate the machine tools with the light off

(B) Machine tools operate without anyone present

(C) A factory needs no workers

Your answer: ________

Task 7 Find the missing words for the following passage and then read it aloud.

Some machinists, often called production ______, may produce large quantities of one _____, especially parts requiring the use of complex ____________ and great ____________. Many modern machine tools are ______(CNC). Frequently, machinists work with computer-control __________ to determine how the automated equipment will cut a part. The programmer may determine the______ of the cut, while the machinist determines the ________ of the cutting tool, the _______ of the cutting tool, and the ________. Because most machinists train in CNC ________, they may write basic programs themselves and often modify ________ in response to problems encountered during test runs. After the production ________ is designed, relatively simple and repetitive operations normally are performed by machine setters, operators, and tenders.

Part B Listening

Task 1 Listen to the five statements twice and write them down.

1. ________________

2. ________________

3. ________________

4. ________________

5. ________________

Task 2 The following video is about machinist job training. Watch it first, then listen to it twice and fill in the blanks with what you hear. Fig. 1-1 shows the Logo of the training center.

This is one of two training centers of Southern California, a great place to become a sought-after （吃香的）(1) __________. Michael Kerwin is the president, "This is a great field to be in." A field which includes grinding, (2) __________, using a (3) __________, and the latest computer-controlled machine tools. The instructors and administrators come right out of the (4) __________. "We were put together by a group of companies and owners of machining companies in this area in

1968 and have been doing training in that area ever since." The training centers with campuses in Norwalk and Ontario are a non-profit arm（部门）of the Los Angeles Chapter of the NTMA. That's the National (5)__________ & (6)__________ Association.

Fig.1-1 Logo of NTMA

Task 3 Watch and listen to the above video once more, and choose the best answer to each of the following questions.

1. How many training centers are there in Southern California?

(A) One. (B) Two. (C) Three.

Your answer: __________

2. The NTMA training centers:

(A) make a lot of money.

(B) are non-profit organizations.

(C) only train men.

Your answer: __________

3. How long is the history of the training center?

(A) More than 20 years. (B) More than 30 years. (C) More than 40 years.

Your answer: __________

4. How long does the basic machining course last?

(A) 700 hours. (B) Seven months. (C) Six weeks.

Your answer: __________

5. What's quite a resume for getting a job of machinists according to what you hear?

(A) Certificate of Entry Level Machinists.

(B) Collection of projects made by the trainees.

(C) A finely made application letter.

Your answer: __________

Part C Speaking

Task 1 Watch the slides and give a name or phrase for each of the slides. You may compete by group. Take notes.

1. ____________________ 6. ____________________

2. ____________________ 7. ____________________

3. ____________________ 8. ____________________

4. ____________________ 9. ____________________

5. ____________________ 10. ____________________

Task 2 As a machinist, you may experience the following English-related working situations. Match them with the corresponding pictures shown in Fig.1-2. Then discuss what other possible situations there are where you might have to use English.

Operate machines with display screens in English ☐

Visit foreign websites for technical information ☐

Read blueprints in English from foreign customers ☐

Follow safety precautions or maintenance practice in English ☐

Look up the machine operator's manual or service manual ☐

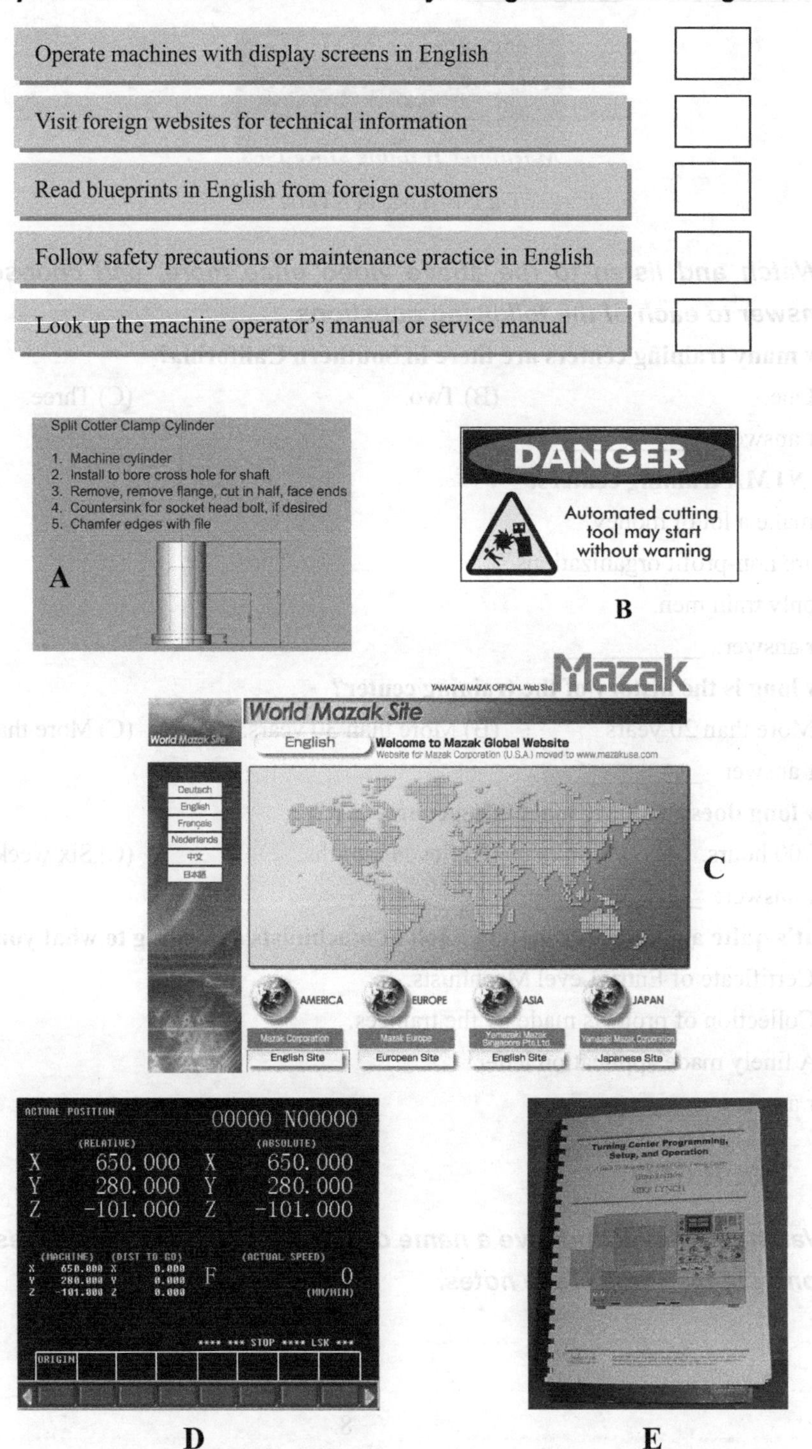

Fig. 1-2 English-related working situations

Task 3 Work in pairs. Take turns with your partner asking and anwering 5 or more questions. You may ask how, what, when, where, why, etc. Two questions have been given for examples.

1. Can you tell me what machinists do?

2. In order to plan and perform the operations needed to make machined products, what knowledge do you think is necessary?

3. ______

4. ______

5. ______

Task 4 Work in groups. Give a PPT presentation in the class. Introduce yourself, and then talk about your major and your future work.

【引导文 1】英文自我介绍样例 Personal introduction

Good morning, my name is Xu Peng. It is really a great honor to have this opportunity to speak here. I am 21 years old, born in Jiangsu Province, and I am currently a senior student at Changzhou Institute of Mechatronic Technology. My major is mechanical manufacturing and automation. I have acquired basic knowledge of mechanical manufacturing both in theory and in practice. I realized the importance of English and began to study diligently when I was eighteen. I hope to pass College English Test 6. This is my greatest wish at the moment.

【引导文 2】英文演示引导句 Presentation

- *开场白*:

Good afternoon, ladies and gentlemen. 女士们先生们，下午好！

On behalf of our company I'd like to welcome you here to ... 我代表本公司欢迎你们光临……

Thank you all for coming here. 谢谢大家来到这里！

Let me introduce myself. I'm Ulrike Huber, Manager's assistant ... 让我自我介绍一下，我叫 Ulrike Huber，是经理助理……

I am a consultant to ... 我是……顾问。

- *引出主题*:

The subject of today's presentation is ... 今天汇报的主题是……

I'll give you an overview of ... 我将给大家谈谈关于……

Today I want to update you on ... 今天我想就……方面更新大家的看法。

● *内容提要：*

Let me first give you a brief overview. 首先请允许我简单介绍一下主要内容。

I'll start off by explaining ..., then focus on ... 我从解释……开始，然后重点……

I'll be talking about ... first, then move on to ... 我先谈谈……，然后谈谈……

● *介绍用例句：*

I'm here today to tell you about ... 今天我要向各位汇报的是……

I've divided my presentation into three sections ... 我的报告分成三部分……

Firstly, I'm going to talk about ... 首先，我来谈谈……

Then I'll give you ... 然后，我给大家……

Finally, I'd like to tell you about ... 最后，我想告诉大家的是……

My presentation will take about five minutes ... 我的报告大约需要 5 分钟……

After that there'll be time for your questions ... 汇报完后，请大家提问……

● *演示用例句：*

I'd like to begin by telling you about ... 我想从……讲起。

Let's move on to ... 我们继续谈下一个问题……

If you look at this diagram, you can see ... 大家看这张图，可以看到……

That brings me to ... 那把我带到了……

To sum up then, ... 总之……

I'd like to conclude by saying ... 我想用……来总结。

Thank you very much for your attention. 谢谢大家！

If you have any questions, I'll be happy to answer them now. 如果有什么问题，现在我很乐意回答。

【引导文 3】切削机械工（机械制造与自动化专业人才）培养方案 Training program for machinists (Diploma of Mechanical Manufacturing and Automation)

相关课程 Related course	课程内容 Course description	技能、知识、能力需求 Skills, knowledge, and abilities needed
数学 Math	● 大学代数 College Algebra ● 几何 Geometry ● 三角学 Trigonometry ● 微积分 Calculus	● 具有数学应用能力 Math skills
工程力学 Engineering mechanics	● 运动 Motions ● 力和力矩 Force and torque ● 功和能 Work and energy ● Capacity and efficiency factor ● 摩擦 Friction ● 材料强度 Strength of materials	● 掌握必需的物理理论知识及应用 Knowledge of physics and the application
计算机应用 Computer application	● Windows ● Word ● Excel ● PowerPoint ● Internet	● 具备基本的计算机应用能力 Basic computer skills
机械制图 Mechanical drafting	● 读图 Blueprint reading ● 机械制图 Mechanical drafting ● 公差与配合 Tolerances and fittings ● 计算机绘图 AutoCAD	● 具有阅读、理解技术图纸和标准的能力 Ability with reading and interpreting technical drawings and standards

续表

相关课程 Related course	课程内容 Course description	技能、知识、能力需求 Skills, knowledge, and abilities needed
机械制造基础 Mechanical manufacturing basics	● 材料学 Materials science - 材料特性 Characteristics of material - 钢和铁材料 Steel and ferrus materials - 非铁金属 Non ferrus metals - 热处理 Heat treatment - 材料测试 Material testing ● 手动加工 Cutting by hand - 划线 Marking - 錾削 Chasing - 锯 Sawing - 锉 Filing ● 机加工 Cutting by machines - 切削用量 Actions and machine variable - 装备和夹具 Apparatus and clamping elements - 钻 Drilling - 车 Turning - 铣 Milling - 磨 Grinding	● 掌握机械工程材料及金属热加工的基本知识 Knowledge of mechanical engineering material and metal heat working ● 制造工艺和材料的知识 Knowledge of production processes, and material ● 普通机加工工艺知识和应用 Knowledge of tooling capabilities and applications for manual machining processes ● 安全标准和防护知识 Knowledge of safety standards and safe guards ● 制造公差范围内的零件的能力 Ability to produce parts within tolerances ● 具有零件测量的能力 Measurement skills
机械制造工艺与装备 Mechanical manufacturing process and apparatus	● 加工工艺 Machining process ● 加工精度 Machining accuracy ● 夹具 Fixtures ● 装配 Assembling	● 掌握机械加工和装配的常规工艺知识，以及一定的夹具和模具设计知识 Knowledge of machining and assembling processes, and fixtures and mould design ● 具有编制与实施机械加工工艺规程和产品装配工艺规程的能力 Ability wiht planning and performing the machining operation and products assembling process ● 具有设计工艺装备的基本能力 Ability with designing process apparatus and fixtures ● 理解并记录质量控制标准 Knowledge of interpreting and documenting quality control standards
液压与气动技术 Hydraulic and pneumatic drive technology	● 液压传动 Hydraulic transmission ● 液压油 Hydraulic oil ● 泵 Pumps ● 液压缸 Hydraulic cylinders ● 控制阀 Control valves ● 液压回路 Hydraulic circuits ● 气动回路 Pneumatic circuits	● 掌握液压、气动技术在产品及装备中的应用技术知识 Knowledge of hydraulic and pneumatic technology in products and equipment application
电气控制与 PLC Electrical control and PLC	● 电流类型/电磁 Kinds of current/magnetism - 直流 Direct current - 交流 Alternating current - 电磁 Electric magnetism ● 电量测量 Measuring of electric variables - 电功率 Electric power - 电流 Current - 电阻 Resistor - 电压 Electric voltage ● 车床的电气控制 Electrical control of lathe ● 铣床的电气控制 Electrical control of mill ● 可编程控制器 PLC	● 具有应用电气控制技术和 PLC 技术的初步能力 Ability with electrical control and PLC application

续表

相关课程 Related course	课程内容 Course description	技能、知识、能力需求 Skills, knowledge, and abilities needed
数控编程 NC programming	• 数控机床特征 Characteristics of CNC machine • 坐标系 Coordinate system • 零点 Zero positions • 数控系统类型 Kinds of control • 创建数控程序 Creation of NC program • 固定循环和子程序 Canned cycle and subprograms • 数控编程-车床 Programming of NC - turning machines • 数控编程-铣床 Programming of NC - milling machines	• 掌握数控机加工工艺知识 Knowledge of tooling capabilities and applications for CNC machining processes • 具有编写数控加工程序及操作的能力 Ability with programming and operation of CNC machine tools • 具有同时监控多台机床的能力 Ability to monitor multiple machines at the same time
CAD/CAM 应用 CAD/CAM application	• 二维设计 2D designing • 三维造型 3D modeling • 数控编程 NC programming	• 具有 CAD/CAM 软件的基本应用能力 Ability with CAD/CAM software application

TASKS

Overview of your PPT

Part D Grammar and Translation

专业英语的特点

专业英语具有下列 5 个特点。

1. 专业词汇多

有些英语词汇在普通英语里和科技英语里的含义在表达时差别很大。

The **overrides** give you the ability to alter the programmed feed and speed, spindle direction, and rapid traverse motion. (Ref. LS 5) 修调（倍率）键用于改变程序中编写的进给速度和主轴转速、主轴转向和快速移动速度。 *override 在普通英语中是“践踏，代理佣金”的意思，而在数控技术中常常指“倍率，修调”。*

又如，**apron** 在普通英语中是“围裙”的意思，而在车床上翻译成“溜板箱”；**engine lathe** 就是指普通车床；**pocket** 有时是“刀套”的意思，有时是“槽，凹处”的意思。这类专业词汇很多，只有大量阅读本专业文献，才能很好地掌握。

2. 被动语态多

科技英语中大量使用被动语态，这是因为文章需要客观地叙述事理，而不是强调动作的主体。为了强调所论述的客观事物，常把它放在句子的首位。

After the layout work **is completed**, machinists perform the necessary machining operations. 规划工作完成以后，机械工就进行必需的加工操作。

Machinists also ensure that the workpiece **is being properly lubricated and cooled**, because the machining of metal products generates a significant amount of heat. 机械工同时要保证工件被恰当润滑和冷却，因为金属产品加工产生大量切削热。

Dull cutting tools **are removed and replaced**. 用钝了的刀具要卸下并更换。

3. 定语（从句）多

科技英语中经常需要说明、定义或限制一些概念、条件等，此时须用定语从句或复杂的限定语来表达。

Lathe is a machine **that turns a piece of metal round and round against a sharp tool that gives it shape**. 车床是一种用尖锐刀具切削旋转金属件的机床，这种尖锐的刀具使金属件获得所要的形状。*在定语从句中还套着一个定语从句 that gives it shape。*

Some machinists, **often called production machinists**, may produce large quantities of one part, especially parts **requiring the use of complex operations and great precision**. 一些机械工（经常称为制造机械工）可能要大批量地制造某种零件，尤其是那些操作复杂和精度要求高的零件。*句中用了过去分词和现在分词作为定语。*

To repair a broken part, maintenance machinists may refer to blueprints and perform the same machining operations **needed to create the original part**. 为了修理已损坏的零件，维修机械工要参考图纸，进行与制造新零件所需的相同的机加工操作。

4．非谓语动词多

英语的每个简单句中，只能用一个谓语动词；如果有几个动词，就必须选出主要动词当谓语，而将其余动作用非谓语动词形式（*v*.-ing, *v*.-ed, to *v*.三种形式）表示，才能符合英语的语法要求。

There is a lot of manual intervention **required to use** a drill press **to drill** holes. 使用台式钻床钻孔，需要很多人工的干预。*这里 required 用过去分词作为定语，to use, to drill 都是非谓语动词形式描述动作。*又如：

They use their knowledge of the working properties of metals and their skill with machine tools **to plan and carry out** the operations **needed to make** machined products that meet precise specifications. *见 NOTES 2。*

Many computer-controlled machines are partially or totally enclosed, **minimizing** the exposure of workers to noise, debris, and the lubricants **used to cool** workpieces during machining. 很多数控机床是全防护或半防护的，最大程度上减少了工人暴露于噪声、切屑碎片和工件冷却润滑液的可能性。

5．复杂长句多

科技文章要求叙述准确，用词严谨，因此一句话里常常包含多个分句，这种复杂且长的句子居科技英语难点之首，阅读翻译时要按汉语习惯加以分析，以短代长，化难为易。

One production machinist, working 8 hours a day, might monitor equipment, replace worn cutting tools, check the accuracy of parts being produced, and perform other tasks on several CNC machines that operate 24 hours a day (lights-out manufacturing). 制造机械工，一天工作八小时，要监控设备运行，更换用钝的刀具，检查被加工零件的精度，同时在几台 24 小时连续运行（无人值守制造）的数控机床上完成其他工作任务。

The headstock is required to be made as robust as possible due to the cutting forces involved, **which can distort a lightly built housing, and induce harmonic vibrations that will transfer through to the workpiece,** reducing the quality of the finished workpiece. (Ref. LS 2) *见 LS2 之 NOTES 2。*

Task　Translate the following sentences or passage into Chinese.

【提示】首先找出每个句子中的主要动词，再分析其他动词的作用。

1．Some machinists, often called production machinists, may produce large quantities of one part, especially parts requiring the use of complex operations and great precision.

……………………………………………………………………………

2．Some manufacturing techniques employ automated parts loaders, automatic tool changers, and computer controls, allowing machine tools to operate without anyone present.

……………………………………………………………………………

3．To repair a broken part, maintenance machinists may refer to blueprints and perform the same machining operations.

……………………………………………………………………………

4. Some rare but increasingly popular metals, such as titanium, are machined at extremely high temperatures.

----

5. They use their knowledge of the working properties of metals and their skill with machine tools to plan and carry out the operations.

----

6. Machine setters, or setup workers, prepare the machines prior to production, perform initial test runs producing a part, and may adjust and make minor repairs to the machinery during its operation. Machine operators and tenders primarily monitor the machinery during its operation; sometimes they load or unload the machine or make minor adjustments to the controls. Because the setup process requires an understanding of the entire production process, setters usually have more training and are more highly skilled than those who simply operate or tend machinery.Many workers both set up and operate equipment.

Learning Situation 

Understand the lathe

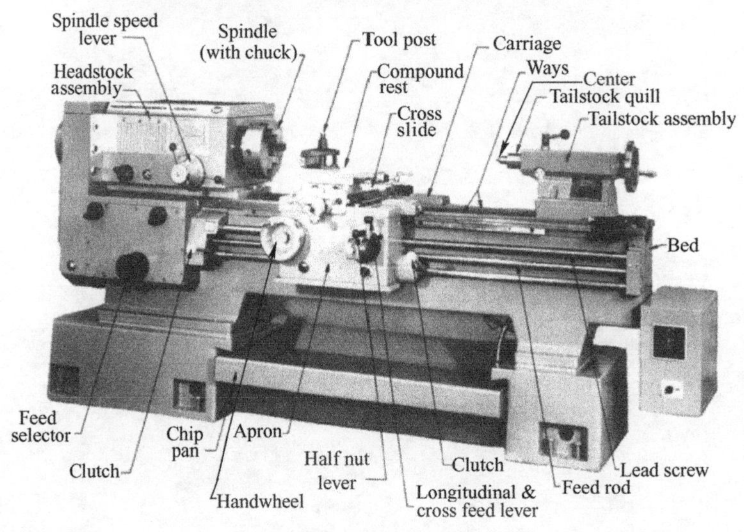

Focus of the situation

This class will introduce you to the components and features of the engine lathe. [本课介绍普通车床的部件和功能。]

Field work

Search the Internet for information about the engine lathe.

Part A Reading

The lathe or engine lathe, as the horizontal metal-turning machine is commonly called, is considered the father of all other machine tools because many of its fundamental mechanical elements are incorporated into the design of other machine tools[1]. Lathes are used in a wide range of applications, and a broad range of materials. These rigid machine tools remove material from a rotating workpiece via the linear movements of various cutting tools, such as tool bits and drill bits.

To gain a good understanding of the lathe, you will need to know the construction and features of the lathe. The names of the various components are illustrated on page 22.

Generally lathes consist of the headstock, the bed, the carriage and the tailstock.

Headstock, spindle and chuck

The headstock comprises the rectangular metal casting at the left end of the lathe. It contains the spindle shaft and its support bearings and the speed change mechanism. See Fig.2-1. The headstock is required to be made as robust as possible due to the cutting forces involved, which can distort a lightly built housing, and induce harmonic vibrations that will transfer through to the workpiece, reducing the quality of the finished workpiece[2].

The spindle is the main rotating shaft on which the chuck is mounted, meaning that the workpiece revolves with it. It is supported by precision thrust bearings mounted in the headstock casting and is driven by an electric motor using a gear box[3]. The spindle is generally hollow to allow long bars to extend through to the work area. This reduces preparation and waste of material.

The spindle speed lever in the front of the headstock changes the gears within the headstock to change the speed and torque of the spindle. The clutch controls the direction of spindle rotation.

Chuck is the clamping device for holding work in the lathe. Three-jaw chuck, four-jaw chuck and six-jaw chuck can be found in the machine shop.

Fig.2-1 Headstock, spindle, and chuck

Bed and ways

The bed is the main body of the lathe, made from rough but sturdy cast iron. The ways are the ground surfaces on the top side of the bed on which the carriage and tailstock ride. See Fig.2-2.

Types of beds include inverted "V" beds, flat beds, and combination "V" and flat beds. "V" and combination beds are used for precision and light duty work, while flat beds are used for heavy duty work.

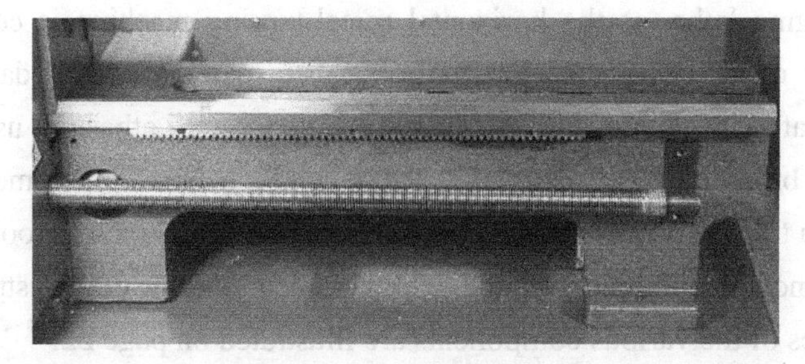

Fig.2-2 Bed and ways

Carriage, apron and saddle

The carriage supports the cross slide, compound rest and tool post and moves along the ways under manual or power feed. It comprises the apron and the saddle. See Fig.2-3.

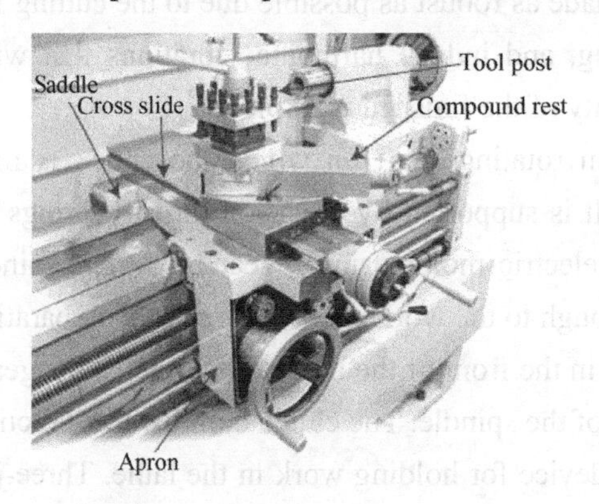

Fig.2-3 Carriage, apron and saddle

The apron is attached to the front of the carriage. It contains the mechanism that controls the movement of the carriage for longitudinal feed and thread cutting. It controls the transverse movement of the cross slide. In general, a lathe apron contains the following mechanical parts:

(a) A carriage handwheel for moving the carriage *by hand* along the bed. This handwheel turns a pinion that engages with a rack gear that is secured to the lathe bed[4]. For short cuts, it is often more convenient to just use the carriage handwheel to advance the carriage, but it is hard to get a nice even finish like you get with power feed.

(b) Gear trains driven by the feed rod. These gear trains transmit power from the feed rod to move the carriage along the ways and to move the cross slide across the ways, thus providing *powered* longitudinal feed and cross feed.

(c) A selective feed lever, here called the longitudinal & cross feed lever, is provided for engaging the longitudinal feed or cross feed as desired.

(d) Friction clutches operated by the longitudinal & cross feed lever on the apron are used to engage or disengage the power feed mechanism[5]. Some lathes have a separate clutch for longitudinal feed and cross feed; others have a single clutch for both.

(e) A selective lever for opening or closing the half-nuts, called half-nut lever.

(f) Half-nuts for engaging and disengaging the lead screw when the lathe is used to cut threads. They are opened or closed by the half-nut lever that is located on the right side of the apron. The half-nuts fit the thread of the lead screw which turns then like a bolt in a nut when they are clamped over it[6]. The carriage is then moved by the thread of the lead screw instead of by the gears of the apron feed mechanisms. The half-nuts are engaged only when the lathe is used to cut threads, at which time the feed mechanism must be disengaged. An interlocking device, that prevents the half-nuts and the feed mechanism from engaging at the same time, is usually provided as a safety feature.

Feed rod, lead screw and half-nut lever

The feed rod transmits power to the apron to drive the longitudinal feed and cross feed mechanisms. The feed rod is driven by the spindle through a train of gears, and the ratio of its speed to that of the spindle can be varied by changing gears to produce various rates of feed[7]. The rotating feed rod drives the gears in the apron. These gears in turn drive the longitudinal feed and cross feed mechanisms through friction clutches. The feed rod is for general machining.

And the lead screw is used only for threading. Along its length are accurately cut ACME threads which engage the threads of the half-nuts in the apron when the half-nuts are clamped over it[8]. See Fig.2-4. When the lead screw turns inside the closed half-nuts, the carriage moves along the ways a distance equal to the lead of the thread in each revolution of the lead screw. Since the lead screw is connected to the spindle through a gear train, the lead screw rotates with the spindle. Whenever the half-nuts are engaged, the longitudinal movement of the carriage is directly controlled by the spindle rotation. The cutting tool is moved a definite distance along the work for each revolution of the spindle.

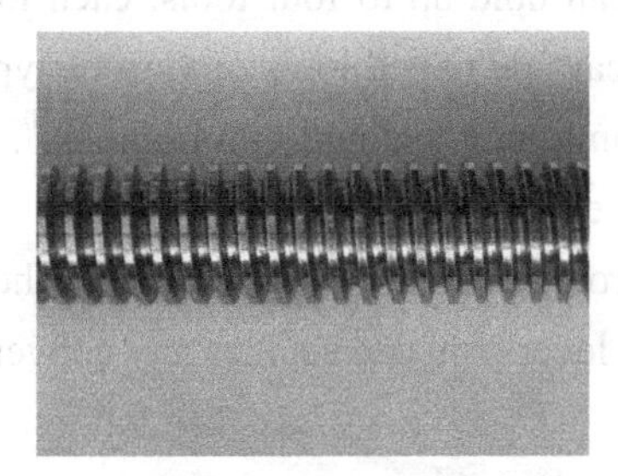

Fig.2-4 Half-nuts and ACME threads

The ratio of the threads per inch of the thread being cut and the threads per inch of the lead screw is the same as the ratio of the speeds of the spindle and the lead screw. For example, if the lead screw and spindle turn at the same speed, the number of threads per inch of the thread being cut is the same as the number of threads per inch of the lead screw. If the spindle turns twice as fast as the lead screw, the number of threads per inch of the thread being cut is twice the number of threads per inch of the lead screw.

Any number of threads can be cut by merely changing the gears in the connecting gear train to obtain the desired ratio of the spindle and the lead screw speeds.

The half-nut lever engages and disengages the power feed to the lead screw.

Cross slide and compound

The cross slide consists of a dovetailed slide that moves at a right angle to the ways. The compound is mounted on the top of the cross slide. The compound bolts into the disk in the cross slide which enables the compound to be rotated. In Fig.2-5 the compound has been removed, showing the cross slide mounted on the saddle.

The cross feed advances the cutting tool into the work at a right angle to the workpiece by means of a lead screw and handwheel. See Fig.2-6.

Fig.2-5 Cross slide mounted on the saddle with compound being removed

Fig.2-6 Underside of the cross slide showing the lead screw

The compound sits on the top of the cross slide and can be rotated to set the tool to advance at an angle to the workpiece[9]. For example, by setting the compound at a 4 degree angle, you can cut shallow tapers. An angle of 30 degrees is sometimes used so that advancing the compound handwheel by .001" actually advances the tool by .0005" (since sine of 30 = .500).

Tool post and tools

The tool post can hold up to four tools, each locked in place by hex-head cap screws. Tools can be ready-made carbide tipped tools of various types or high-speed steel (HSS) ground to shape by the user from commonly available tool blanks[10]. See Fig.2-7. The tool is fed into the rotating workpiece to machine it to size.

The tool post rotates around a large bolt on the top of the compound slide and is locked in place by a clamping lever. You can set it to any convenient angle.

Fig.2-7 Tool post and tool blank

Tailstock and centers

The tailstock casting rides on the ways and moves on the ways along the length of the bed to accommodate work of varying lengths. It can be locked in the desired position by the tailstock

Fig.2-8 Tailstock and center

clamping nut. See Fig.2-8.

The primary purpose of the tailstock is to hold the dead center to support one end of the work being machined between centers. However, it can also be used to hold live centers, tapered shank drills, reamers, and drill chucks.

The dead center is held in a tapered hole in the tailstock quill. The quill is moved back and forth in the tailstock barrel for longitudinal adjustment. The handwheel advances the quill for drilling and tapping operations. The quill is kept from revolving by a key that fits a keyway, cut along the bottom of the quill.

Well, since we are now familiarized with the basic parts of the lathe, let's get started with the machining operations.

TECHNICAL WORDS

horizontal	[ˌhɔriˈzɔntl]	*a.*	卧式的，水平的
mechanical	[miˈkænikl]	*a.*	机械的
rigid	[ˈridʒid]	*a.*	刚性的
rotate	[rəuˈteit]	*v.*	旋转
linear	[ˈliniə]	*a.*	直线的
construction	[kənˈstrʌkʃən]	*n.*	结构；建设
feature	[ˈfiːtʃə]	*n.*	功能，特点，特征
component	[kəmˈpəunənt]	*n.*	部件，元件
headstock	[ˈhedstɔk]	*n.*	床头箱
tailstock	[ˈteilstɔk]	*n.*	尾架
chuck	[tʃʌk]	*n. & v.*	卡盘；夹紧
rectangular	[rekˈtæŋgjulə]	*a.*	矩形的，成直角的
casting	[ˈkɑːstiŋ]	*n.*	铸件
bearing	[ˈbɛəriŋ]	*n.*	轴承
mechanism	[ˈmekənizəm]	*n.*	机构
housing	[ˈhauziŋ]	*n.*	机架，机座
revolve	[riˈvɔlv]	*v.*	旋转
lever	[ˈliːvə, ˈlevə]	*n.*	手柄
torque	[tɔːk]	*n.*	扭矩，转矩
clutch	[klʌtʃ]	*n.*	离合器
grind	[graind]	*v.*	磨削

carriage	[ˈkæridʒ]	*n.*	床鞍，大滑板
apron	[ˈeiprən]	*n.*	溜板箱
saddle	[ˈsædl]	*n.*	滑鞍
longitudinal	[ˌlɔndʒiˈtjuːdinl]	*a.*	纵向的
transverse	[ˈtrænzvəːs]	*a.*	横向的
pinion	[ˈpiniən]	*n.*	小齿轮
rack	[ræk]	*n.*	齿条
engage	[inˈgeidʒ]	*v.*	啮合
secure	[siˈkjuə]	*v.*	固定
even	[ˈiːvən]	*a.*	均匀的，平稳的，平均的
finish	[ˈfiniʃ]	*n. & v.*	表面光洁度（也称 surface finish）；精加工
nut	[nʌt]	*n.*	螺母
bolt	[bɔlt]	*n. & v.*	螺栓；闩上
interlock	[ˌintəˈlɔk]	*n. & v.*	互锁
ratio	[ˈreiʃiəu]	*n.*	比，比率
definite	[ˈdefinit]	*a.*	一定的
taper	[ˈteipə]	*n. & v.*	锥形，锥面；锥面切削
reamer	[ˈriːmə]	*n.*	铰刀
quill	[kwil]	*n.*	钻轴，活动套筒
tap	[tæp]	*v. & n.*	攻丝；丝锥
keyway	[ˈkiːˌwei]	*n.*	键槽

PHRASES

engine lathe	普通车床
turning machine	车床
cutting force	切削力
finished workpiece	成品件
precision thrust bearing	精密止推轴承
gear box	齿轮箱，变速箱
three-jaw chuck	三爪卡盘
cast iron	铸铁
light duty	轻负荷
heavy duty	重载，强力
cross slide	中滑板，横向滑板

compound rest		复式刀架
compound slide	（compound）	小滑板，复式滑板
tool post		（方）刀架
thread cutting	（threading）	螺纹切削
gear train		齿轮系
feed rod		光杠
cross feed		横向进给
friction clutch		摩擦离合器
half-nut		开合螺母，半开螺母
lead screw		丝杠
ACME thread		梯形螺纹
thread of the lead screw	（lead of the thread）	丝杠螺距（导程）
threads per inch	（TPI）	每英寸螺头数
dovetailed slide		燕尾式滑板
right angle		直角
hex-head cap screw		六角螺钉
carbide tipped tool		硬质合金刀具
high-speed steel	（HSS）	高速钢
tool blank		刀坯
dead center		死顶尖
live center		活顶尖
tapered shank drill		锥柄钻头
drill chuck		钻夹

NOTES

1. The lathe or engine lathe, **as the horizontal metal-turning machine is commonly called,** is considered the father of all other machine tools, because many of its fundamental mechanical elements are incorporated into the design of other machine tools. 卧式金属车床常称为车床或普通车床，被认为是机床之父，因为其他机床的设计中都包含了车床的很多基本机械要素。*as the horizontal metal-turning machine is commonly called 为定语从句，关系代词 as 引起定语从句，指代 the lathe or engine lathe，在定语从句中作为宾语补足语。*

2. The headstock is required to be made as robust as possible due to the cutting forces **involved, which can distort a lightly built housing, and induce harmonic vibrations that will transfer through to the workpiece**, reducing the quality of the finished workpiece. 床头箱的制造必须十分坚固，因为切削力的影响可能会使制造不坚固的机座变形，并且产生谐振，而谐振传至工件，会降低成品件的质量。*这是一句典型的复杂长句，which 引导定语从句，对 cutting*

forces 进一步说明，involved 是过去分词作为后置定语，修饰cutting forces；harmonic vibrations that will transfer through to the workpiece 是定语从句中的定语从句。reducing the quality of the finished workpiece 是现在分词作为非谓语动词，对切削力产生的影响进行补充说明。

3. It is supported by precision thrust bearings **mounted in the headstock casting** and is driven by an electric motor using a gear box. 主轴由安装在床头箱铸件中的精密止推轴承支撑，采用齿轮箱通过电动机驱动。*这里 mounted in the headstock casting 是过去分词作为定语，修饰 precision thrust bearings，using a gear box 是非谓语动词形式。*

4. This handwheel turns a pinion **that engages with a rack gear that is secured to the lathe bed**. 这个手轮转动齿轮，使之与固定在床身上的齿条啮合。*that engages with a rack gear 为定语从句，修饰pinion，that is secured to the lathe bed 为定语从句中的定语从句，修饰rack gear。*

5. Friction clutches **operated by the longitudinal & cross feed lever on the apron** are used to engage or disengage the power feed mechanism. 由溜板箱上的纵向/横向进给手柄操作的摩擦离合器用于啮合或断开自动进给机构。*operated by...是过去分词作为定语，修饰 friction clutches.*

6. The half-nuts fit the thread of the lead screw **which turns then like a bolt in a nut** when they are clamped over it. 开合螺母夹住丝杠时，开合螺母合上丝杠的螺纹，然后丝杠像螺栓一样在螺母中转动。

7. The feed rod is driven by the spindle through a train of gears, and the ratio of **its** speed to **that** of the spindle can be varied by changing gears to produce various rates of feed. 进给光杠通过齿轮系由主轴驱动，光杠速度与主轴速度的比率可通过切换齿轮改变，以产生不同的进给速度。*这里两个代词its 和that，前者指feed rod，后者指speed。*

8. And the lead screw is used only for threading. Along its length are accurately cut ACME threads which engage the threads of the half-nuts in the apron when the half-nuts are clamped over it. 丝杠仅用于螺纹切削，其长度方向为精密切削的 ACME 螺纹，当开合螺母夹住丝杠时，与溜板箱内开合螺母的螺纹啮合。*ACME 螺纹为英制梯形螺纹，常用于手动机床。*

9. The compound sits on the top of the cross slide and can be rotated to set the tool to advance at an angle to the workpiece. 小滑板坐在中滑板的上面，可以旋转来设定刀具以某个角度向工件进刀。

10. Tools can be ready-made carbide tipped tools of various types or high speed steel (HSS) **ground** to shape by the user from commonly available tool blanks. 刀具可以是各种现成的硬质合金刀尖刀具，也可以由用户用常见的刀坯磨成不同刀尖形状的高速钢刀具。*这里 ground to shape by...是过去分词作为后置定语，相当于which is ground to shape by...。*

PRACTICE

Task 1　Translate the following words and phrases into English.

1．应用广泛	2．直线运动
3．各种刀具	4．变速机构
5．谐振	6．成品件

7. 精密止推轴承
8. 齿轮切换
9. 磨削平面
10. 强力切削
11. 纵向进给
12. 横向进给
13. 齿轮齿条机构
14. 摩擦离合器
15. 丝杠
16. 光杠
17. 互锁装置
18. 直角
19. 大滑板
20. 中滑板
21. 小滑板
22. 尾架套筒

Task 2 Fill in the brackets with words that have similar meaning to the underlined words, changing their forms if necessary.

1. ( ) Lathes are used in a wide range of applications, and a broad range of materials.

2. ( ) The names of the various components are illustrated on the previous page.

3. ( ) The carriage comprises the apron and the saddle.

4. ( ) The feed rod is driven by the spindle through a train of gears, and the ratio of its speed to that of the spindle can be varied by changing gears to produce various rates of feed.

5. ( ) The spindle is the main rotating shaft on which the chuck is mounted, meaning that the workpiece revolves with it.

6. ( ) The half-nuts fit the thread of the lead screw which turns then like a bolt in a nut when they are clamped over it.

7. ( ) The spindle is supported by precision thrust bearings mounted in the headstock casting and is driven by an electric motor using a gear box.

8. ( ) The cross slide consists of a dovetailed slide that moves at a right angle to the ways.

9. ( ) To gain a good understanding of the lathe, you will need to know the construction and features of the lathe.

10. ( ) This handwheel turns a pinion that engages with a rack gear that is secured to the lathe bed.

Task 3 Complete the sentences with proper words or phrases. Some have been given in the word bank and you just make selection, changing their forms if necessary.

cross slide, carriage, engine lathe, cutting tool, spindle, half-nut, workpiece

1. The ____________, as the horizontal metal-turning machine is commonly called, is usually considered the father of all other machine tools.

2. Lathes remove material from a rotating ____________ via the linear movements of various ____________, such as tool bits and drill bits.

3. Generally lathes consist of the headstock, the bed, the ____________ and the tailstock.

4. The workpiece revolves with the ____________on a lathe.

5. ______________are provided for engaging and disengaging the lead screw when the lathe is used to cut threads.

6. The compound sits on the top of the ___________ and can be rotated to set the tool to advance at an angle to the workpiece.

7. When the lead screw turns inside the closed half-nuts, the carriage moves along the ways a distance equal to the ___________ of the thread in each revolution of the lead screw.

8. An angle of 30 degrees is sometimes used so that advancing the compound handwheel by .001" actually advances the tool by___________.

9. Movable parts on a lathe are ____.

10. The carriage handwheel turns a ___________ that engages with a rack gear that is secured to the lathe bed.

Task 4 Decide whether the following statements are true (T) or false (F).

1. (   ) The half-nuts are engaged only when the lathe is used to cut threads, at which time the feed mechanism must be disengaged.

2. (   ) Lathes remove material from a rotating cutting tool via the linear movements of workpiece.

3. (   ) The lathe spindle is solid.

4. (   ) To change the speed and torque of the spindle, we can operate the spindle speed lever on the apron.

5. (   ) The half-nut lever is located on the apron.

6. (   ) Lathes can only be used for machining components made of metal.

Task 5 Choose the best answer.

1. Which machine is the father of all machine tools?

(A) The mill. (B) The lathe. (C) The drill.

Your answer:__________

2. What's the purpose of the compound?

(A) To hold the center. (B) To cut threads. (C) To cut tapers.

Your answer: __________

3. The machine that turns a piece of metal round and round against a sharp tool is:

(A) Milling machine. (B) Lathe. (C) Boring machine.

Your answer:__________

4. The lathe is widely used in producing___________work.

(A) Box-type (B) Round (C) Odd-shaped

Your answer:__________

5. What is the lathe bed made of?

(A) Aluminum. (B) Steel. (C) Cast iron.

Your answer:__________

6. Which component supports and rotates workpieces about the axis of the lathe?

(A) Spindle. (B) Quill. (C) Carriage.

Your answer:__________

7. What part of the lathe provides a base for the working parts?

(A) Carriage and apron. (B) Bed and ways. (C) Headstock and spindle.

Your answer:__________

8. When cutting threads, what components in the lathe apron engage and disengage the lead screw?

(A) Half-nuts. (B) Gear trains. (C) Rack and pinion.

Your answer:__________

9. For heavy duty work, __________beds are used.

(A) Inverted "V" (B) Flat (C) Combination "V" and flat

Your answer: __________

10. The power feed is engaged by the____________________on the apron.

(A) Spindle speed lever

(B) Half-nut lever

(C) Longitudinal& cross feed lever

Your answer: __________

Task 6 Fill in the blanks with words or phrases from the reading that match the meanings in the column on the right. The first letters are already given. Then compare with your partner.

1. a__________ Front part of the carriage assembly on which the carriage handwheel is mounted.

2. b__________ Main supporting casting running the length of the lathe.

3. c__________ Assembly that moves the tool post and cutting tool along the ways.

4. carriage h_______ A wheel with a handle used to move the carriage by hand by means of a rack and pinion drive.

5. c__________ A clamping device for holding work in the lathe or for holding drills in the tailstock.

6. f__________ A lathe operation in which metal is removed from the end of workpiece to create a smooth perpendicular（垂直的）surface, or face.

7. h__________ The main casting mounted on the left end of the bed, in which the spindle is mounted. Houses the spindle speed change gears.

8. s__________ A casting, shaped like an "H" when viewed from above, which rides along the ways. Along with the apron, it is one of the two main components that make up the carriage.

9. s__________ Main rotating shaft on which the chuck or other work holding device is mounted. It is mounted in precision bearings and passes through the headstock.

10. t__________ Cast iron assembly that can slide along the ways and be locked in place. Used to hold long work in place or to mount a drill chuck for drilling into the end of the work.

11. tool p________ A holding device mounted on the compound into which the cutting tool is clamped.

12. t___________ A lathe operation in which metal is removed from the outside diameter of the workpiece, thus reducing its diameter to a desired size.

13. d___________ A multi-point cutting tool used to make round holes.

14. w__________ Two parallel（平行的）machined rails that are the ground surfaces on the top side of the bed.

15. c__________ Movable platform on which the tool post is mounted; can be set at an angle to the workpiece.

16. A__________ The most common type of lead screw found in machine tool applications. Compared to a ball screw, these lead screws have very high friction and backlash（反向间隙）.

Task 7 Write the English name of each component shown in Fig.2-9 on the corresponding number. Some have been given.

1----- 2a-----

2----- 2b-----

3----- 3b-----

4----- 5a-----

5----- 5b----half-nut lever

5c----carriage power feed mechanism

Fig.2-9

Task 8 Find the missing words for the following passage and then read it aloud.

The headstock comprises the rectangular metal________at the left end of the lathe. It contains the spindle shaft and its support _________ and the speed change mechanism. The headstock is required to be made as robust as possible due to the cutting________ involved, which can distort a lightly built ______________, and induce harmonic vibrations that will transfer through to the ___________, reducing the quality of the finished workpiece.

The spindle is the main rotating _______ on which the chuck is mounted, meaning that the workpiece revolves with it. It is supported by precision thrust bearings mounted in the headstock casting and is driven by an electric ________ using a gear box. The spindle is generally hollow to allow long________ to extend through to the work area. This reduces preparation and waste of material.

The spindle speed lever in the front of the headstock changes the gears within the headstock to change the _____ and_______ of the spindle. The clutch controls the direction of spindle rotation.

Part B Listening

Task 1 Listen to the five statements twice and write them down.

1. ....
2. ....
3. ....
4. ....
5. ....

Task 2 The following video clip is about lathe parts. Watch it first, then listen to it twice and complete the following table according to what you hear and see.

No.	Name of the parts	No.	Name of the parts
1		8	
2		9	
3		10	
4		11	
5		12	
6		13	
7		14	

Task 3 The following video clip is about lathe motions. Watch it first, then listen to it twice and fill in the following blanks with what you hear and see.

1. The spindle________________in both_________________.
2. The tailstock can _________along_______ and the quill can_______within the tailstock.
3. The carriage also_____________________________.
4. The cross slide____________perpendicular to_________.
5. The compound slide__________at any preset angle over______________.
6. The lead screw and feed rod also___________ and reserve for certain automatic functions.

Task 4 The following video introduces the oldest machine tool in the machine shop. Watch it first, then listen to it twice and fill in the following blanks with what you hear and see.

Throughout the past century (1)________ operated lathes have been the mainstay（支柱）of tool rooms and small machine shops everywhere. These machines, however, take considerable time to(2) ______________ and are quite cumbersome（麻烦的）to operate. In fact, the setup time often far(3) ______________ the actual time it takes to make the part. And while many of these machines have digital readouts for accurate(4) ______________, there is no way to have a manual machine repeat a process (5) ______________ to make multiple parts from the same set of movements.

Part C Speaking

Task 1 Watch the slides and give the English name or description for each of the slides. Take notes.

1. ______________ 6. ______________
2. ______________ 7. ______________
3. ______________ 8. ______________
4. ______________ 9. ______________
5. ______________ 10. ______________

Task 2 Work in pairs. Take turns with your partner asking and answering 5 or more questions. Two questions have been given for examples.

1. What are the uses of the lathe?
2. Why is the spindle generally hollow?
3. ______________
4. ______________
5. ______________

Task 3 Work in pairs. Explain the following technical words and phrases in English.

cross slide, carriage, engine lathe, cutting tool, spindle, half-nut, workpiece

Task 4 Watch the video clip about lathe motions with the loudspeaker mute and tell the class the lathe motions just as a narrator.

Task 5 Work in groups. Create a technical situation about the lathe and make a conversation in your group. Take notes when others are speaking. The group leader chairs the conversation or discussion.

TASKS

Notes

Part D Grammar and Translation

定语的译法

科技英语中，经常需要对某个概念进行定义，这时常使用定语或定语从句，参见学习情境 1 中 Reading 之 Task 3。翻译成汉语时，可将长的定语分为并列的两句或多个单句，对于短的定语则用"……的……"。

1. 过去分词作为定语

过去分词作为定语，通常句意多为被动，后置时相当于定语从句，只是省略了 which。如：

Tools can be ready-made carbide tipped tools of various types or high speed steel (HSS) **ground** to shape by the user from commonly available tool blanks. *见 NOTES 10。*

It is supported by precision thrust bearings **mounted** in the headstock casting and is driven by an electric motor using a gear box. *见 NOTES 3。*

They use their knowledge of the working properties of metals and their skill with machine tools to plan and carry out the operations **needed** to make **machined** products that meet precise specifications. (Ref. LS 1) *见 LS 1 之 NOTES 2。*

The ways are the **ground** surfaces on the top side of the bed on which the carriage and tailstock ride. 导轨是床身上**磨削的**表面，床鞍和尾架跨骑在床身上。*这里 ground 是过去分词作为定语，已形容词化。又如，given cutting speed（给定的切削速度），desired position（期望的位置），tapered hole（锥孔）。*

2. 现在分词作为定语

The primary purpose of the tailstock is to hold the dead center to support one end of the work

being machined between centers. 尾架的主要作用是安装死顶尖以支撑两顶尖之间的**被加工**工件的一端。

The ratio of the threads per inch of the thread **being cut** and the threads per inch of the lead screw is the same as the ratio of the speeds of the spindle and the lead screw. **被切削**螺纹的每英寸螺纹大小和丝杠的每英寸螺纹大小的比率与主轴转速和丝杠转速的比率相同。

Determining the most advantageous feeds and speeds for a particular lathe operation depends on the kind of material **being worked on**, the type of tool, the diameter and length of the workpiece, the type of cut **desired** (rough or finishing), the cutting oil **used**, and the condition of the lathe **being used**. (Ref. LS 3) 确定针对某一特定车床操作的最佳的进给速度和切削速度取决于**被加工的**材料类型、刀具类型、工件的直径和长度、**期望的**切削类型（粗加工还是精加工）、**所使用的**切削油，以及所使用车床的条件。

Any number of threads can be cut by merely changing the gears in the **connecting** gear train to obtain the **desired** ratio of the spindle and the lead screw speeds. 只要变换**所连**齿轮系中的齿轮获得**预期的**主轴和丝杠转速比，就可以切削任意螺距的螺纹。*又如，rotating workpiece（旋转工件），clamping lever（锁紧手柄），work of varying length（变化长度的工件）等。*

3．不定式作为定语

To determine the rotational speed necessary to produce a given cutting speed, it is necessary to know the diameter of the workpiece **to be cut**. (Ref. LS 3) 为了确定产生给定的切削速度所需的转速，有必要知道**待切削**工件的直径。

Since this form of CNC machine can perform multiple operations in a single program (as many CNC machines can), the beginner should also know the basics of how to process workpieces machined by turning so a sequence of machining operations can be developed for workpieces **to be machined**. (Ref. LS 4) *见 LS 4 之 NOTES 3。*

现在分词作为定语后置多表示正在进行，过去分词作为定语后置多表示已完成的动作或被动句意，而动词不定式作为定语后置表示要做的事尚未发生，常与名词有动宾关系，不及物动词加介词。以上三种定语形式经常翻译为“所……”。

4．形容词作为后置定语

To determine the rotational speed **necessary** to produce a given cutting speed, it is necessary to know the diameter of the workpiece to be cut. (Ref. LS 3) 为了确定产生给定的切削速度所需的转速，有必要知道待切削工件的直径。*rotational speed necessary 相当于 rotational speed which is necessary。*

5．定语从句

The lathe or engine lathe, **as the horizontal metal-turning machine is commonly called**, is usually considered the father of all other machine tools, because many of its fundamental mechanical elements are incorporated into the design of other machine tools. *见 NOTES 1。*

The spindle is the main rotating shaft **on which the chuck is mounted**, meaning that the workpiece revolves with it. 主轴是装有卡盘的主要旋转轴，即工件和主轴一起旋转。*对于较*

短的定语从句，翻译成"……的"。

A lathe is a machine **that turns** a piece of metal round and round against a sharp tool. 车床是一种相对尖锐的刀具连续转动金属材料的机床。*通常定义、解释某事物时用到定语从句。*

This handwheel turns a pinion **that engages** with a rack gear **that is secured** to the lathe bed. *见NOTES 4。*

Task Find the attributes in the following sentences and translate them into Chinese.

Model: The cross slide consists of a dovetailed slide that moves at a right angle to the ways.

This sentence is combined with the following two sentences:

The cross slide consists of a dovetailed slide. The dovetailed slide moves at a right angle to the ways.

1. The tailstock can be used to hold live centers, tapered shank drills, reamers, and drill chucks.

----

2. An interlocking device, that prevents the half-nuts and the feed mechanism from engaging at the same time, is usually provided as a safety feature.

----

3. The compound bolts into the disk in the cross slide which enables the compound to be rotated.

----

4. It can be locked in the desired position by the tailstock clamping nut.

----

5. The cross slide consists of a dovetailed slide that moves at a right angle to the ways.

----

6. One production machinist might monitor equipment, replace worn cutting tools, check the accuracy of parts being produced, and perform other tasks on several CNC machines that operate 24 hours a day.

Learning Situation 

Master basic machining practice

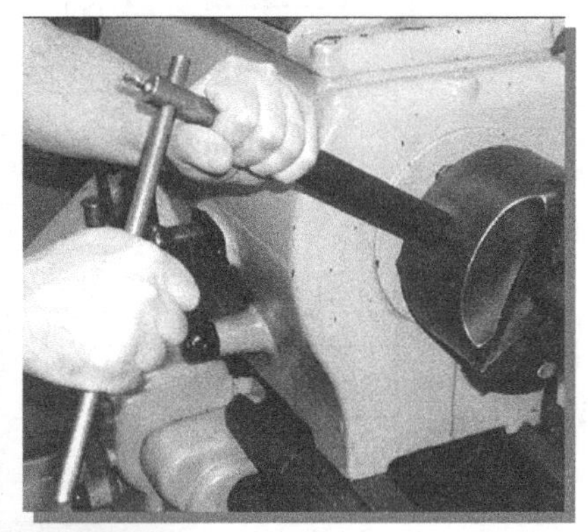

Focus of the situation

This class will introduce you to the cutting data and lathe safety, and teach you the turning operations performed on the engine lathe. [本课介绍切削用量和车床安全的知识，教你普通车床的车削操作。]

Field work

Participate in shop turning practice and try to make the Job Breakdown Sheet.

Part A Reading

General operations on the lathe include turning, facing, boring, drilling, reaming, tapping, grooving, parting, chamfering, tapering, threading, forming, and knurling. See Fig.3-1. Before these operations can be done, a thorough knowledge of the variable factors of lathe speeds, feeds, and depth of cuts must be understood. And good safety practices should be followed to ensure safe machining.

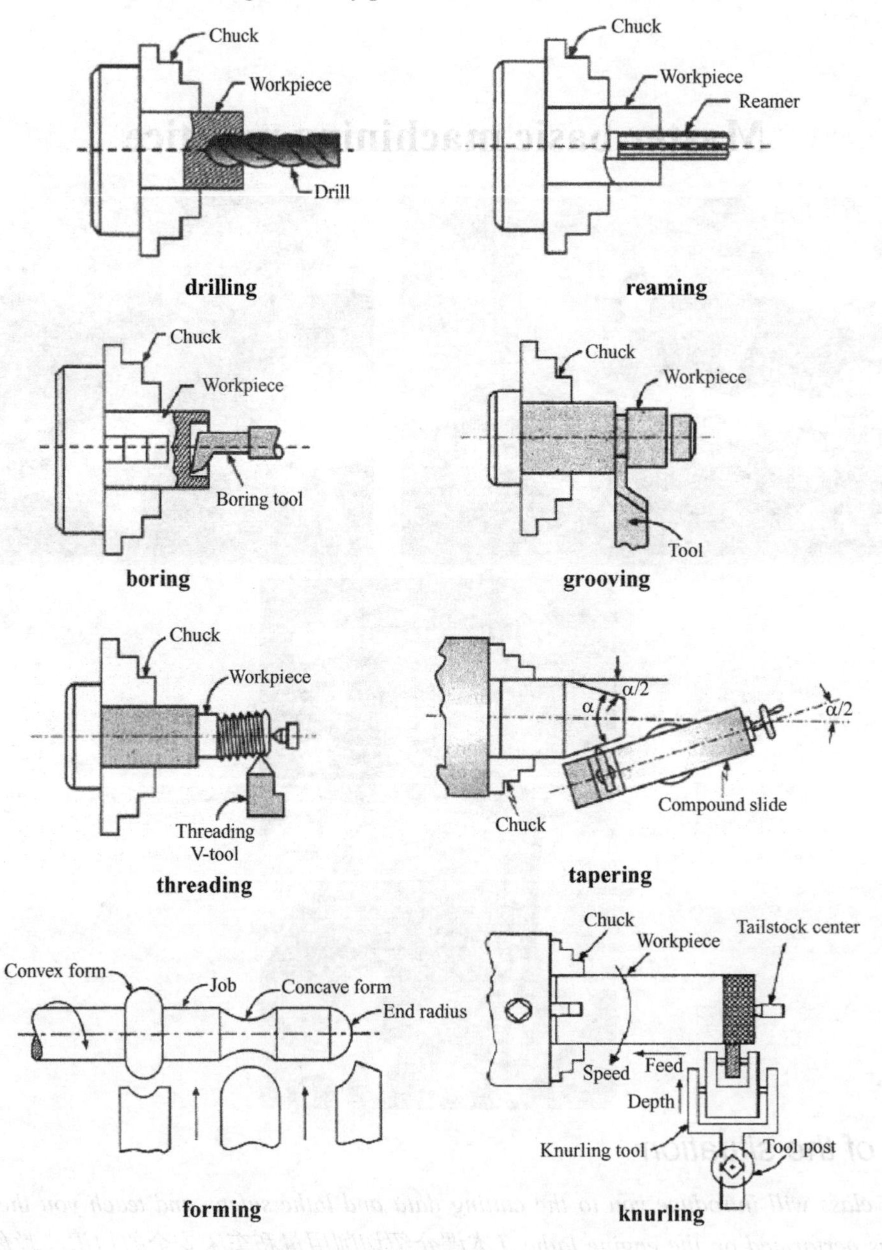

Fig.3-1 Various lathe operations

Three elements

Determining the most advantageous feeds and speeds for a particular lathe operation depends on the kind of material being worked on, the type of tool, the diameter and length of the workpiece, the type of cut desired (rough or finishing), the cutting oil used, and the condition of the lathe being used. The three elements-cutting speed, feed, and depth of cut, are shown in Fig.3-2.

(1) Cutting speed V_C.

The cutting speed of a tool bit is defined as the distance of the workpiece surface, measured at the circumference, that passes the tool bit in one minute[1]. The cutting speed, expressed in meters per minute, must not be confused with the spindle speed of the lathe which is expressed in revolutions per minute. See Fig.3-3. To obtain uniform cutting speed, the lathe spindle must be revolved faster for workpieces of small diameters and slower for workpieces of large diameters.

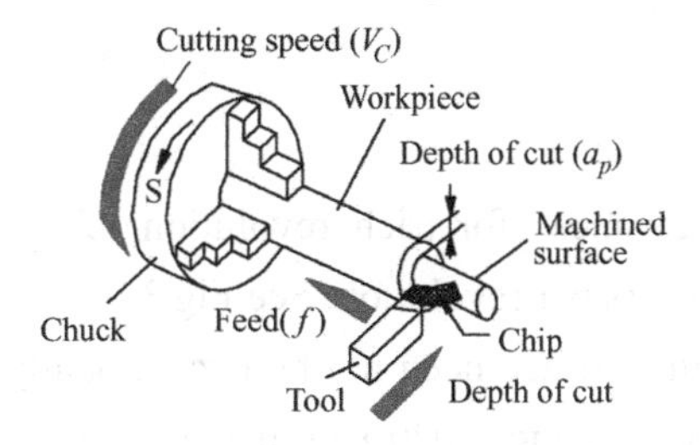

Fig.3-2 Illustration of three elements

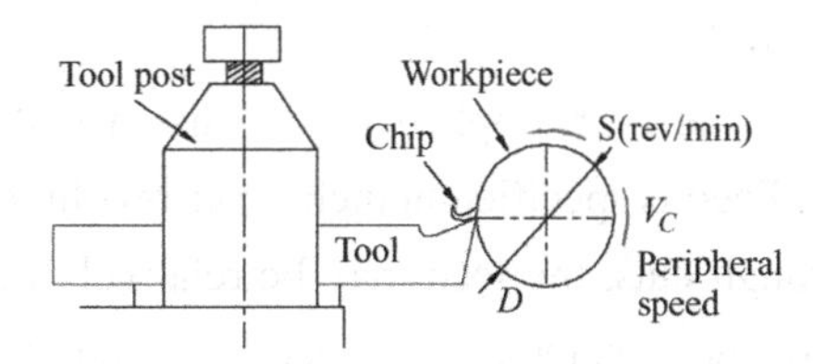

Fig.3-3 Relationship between cutting speed and spindle speed

The proper cutting speed for a given job depends upon the hardness of the material being worked on, the material of the cutter bit, and the feed and depth of cut to be used. Table 3-1 lists specific ranges of cutting speeds for straight turning and for threading under normal conditions. It is proper to start machining operations at these speeds and observe the effect on the cutter bit and workpiece. If the cutter bit does not cut satisfactorily, the speed should be reduced. Carbon steel tools, when used, require a reduction in speed because they cannot withstand the heat produced as a result of high speed turning. Carbide tipped tools, however, will stand speeds in excess of those recommended for high-speed steel tools. The feed and depth of cut should be average as described in paragraphs (2) and (3) below, for the recommended speeds in Table 3-1. If it is desired to increase either the feed or the depth of cut, the cutting speed should be proportionally reduced to prevent overheating and excessive cutter bit wear.

Table 3-1 Lathe cutting speeds for straight turning and threading

Material	Straight turning speed(m/min)	Threading speed(m/min)
Aluminum	60 to 90	15
Brass, yellow	45 to 60	15
Bronze, soft	24 to 30	9
Bronze, hard	9 to24	6
Cast iron	15 to 24	7.5
Copper	18 to 24	7.5
Steel, high carbon (tool)	10.5 to 12	4.5
Steel, low carbon	24 to 30	10.5
Steel, medium carbon	18 to 24	7.5
Steel, stainless	12 to 15	4.5

Note: The speeds are based on the use of high-speed steel cutter bits. These speeds may be increased 25 to 50 percent, if cutting oil is used. If carbide tipped cutter bits are used, speeds may be 2 or 3 times as high as those given for high-speed steel cutter bits. If carbon steel cutter bits are used, the speed should be reduced by about 25 percent.

To determine the rotational speed necessary to produce a given cutting speed, it is necessary to know the diameter of the workpiece to be cut. To calculate the spindle speed, knowing the diameter of the workpiece, use the following formula:

$$S = \frac{1000 \cdot V_C}{\pi \cdot D}$$

Where, S = spindle speed or RPM (r/min)

V_C = cutting speed (m/min)

D = diameter of the workpiece (mm)

(2) Feed f.

Feed is the term applied to the distance the tool bit advances for each revolution of the workpiece. Feed is specified in inches per revolution or millimeters per revolution. See Fig.3-4.

For rough cuts, the feed may be relatively heavy since the surface need not be exceptionally smooth. For most materials, the feed for rough cuts should be 0.010 to 0.020 inch per revolution. For finishing cuts, a light feed is necessary since a heavy feed produces a poor finish. If a large amount of stock is to be removed, it is advisable to take one or more roughing cuts and then take light finishing cuts at relatively high speeds[2].

(3) Depth of cut a_p.

The depth of cut regulates the reduction in the diameter of the workpiece for each longitudinal traverse of the tool bit. The workpiece diameter is reduced by twice the depth of cut in each complete traverse of the tool bit. Generally, the deeper the cut, the slower the speed, since a deep cut requires more power. See Fig.3-5.

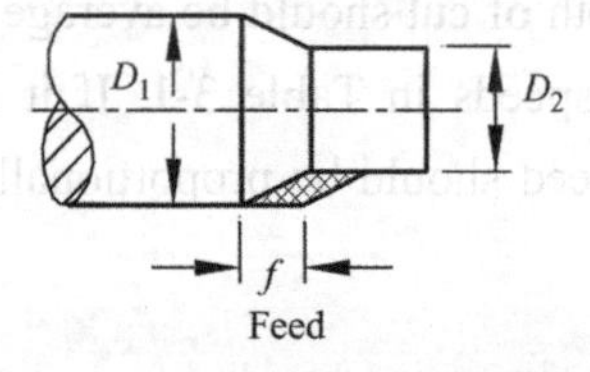

Fig.3-4 Illustration of feed

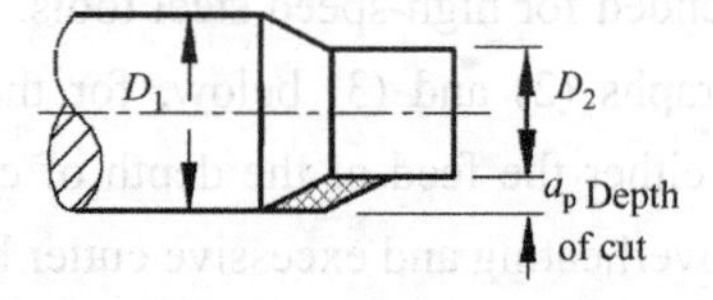

Fig.3-5 Illustration of depth of cut

The depth of cut for roughing is generally five to ten times deeper than the feed. The reason for this is that more of the cutting edge of the tool bit is in contact with the workpiece for the amount of metal being removed[3]. For roughing with feeds of from 0.010 to 0.020 inch per revolution, the depth of cut should be between 3/16 and 1/4 inch. Finishing cuts are generally very light, therefore the cutting speed can be increased since the chip is thin.

Lathe safety precautions

In machining operations, there is one sequence of events that one must always follow: SAFETY FIRST, ACCURACY SECOND, AND SPEED LAST. With this in mind, let's look at some of the more important safety precautions that should be observed before and during lathe operations.

- Always wear safety glasses with side-shields. The lathe can throw off sharp, hot metal chips at considerable speed that can be quite hazardous.
- Wear short sleeve shirts. Roll sleeves above elbows if long shirts. Loose sleeves can catch on rotating work and quickly pull your hand or arm into harm's way.
- Wear leather work shoes to protect your feet from sharp metal chips on the shop floor and from tools and chunks of metal that may get dropped.
- Remove wrist watches, necklaces, chains and other jewelry. It's a good idea even to remove your finger ring since it can catch on rotating work and severely damage your ring finger and hand.
- Tie back long hair so it can't get caught in the rotating work. Think about what happens to your face if your hair gets entangled.
- Always double check to make sure your work is securely clamped in the chuck or between centers before starting the lathe. Start the lathe at low speed and increase the speed gradually.
- Get in the habit of removing the chuck key immediately after use. It is recommended that you never remove your hand from the chuck key when it is in the chuck. The chuck key can be a lethal projectile if the lathe is started with the chuck key in the chuck.
- Handle heavy chucks with care and protect the lathe ways with a block of wood when installing a chuck.
- Know where the emergency stop is before operating the lathe.
- Handle sharp cutters, centers, and drills with care.
- Keep tools overhang as short as possible.
- Keep your fingers clear of the rotating work and cutting tools. Never attempt to break away metal spirals as they form at the cutting tool. Never attempt to measure work while it is turning. Always stop the lathe before making adjustments.
- Avoid reaching over the spinning chuck. For filing operations, hold the tang end of the file in your left hand so that your hand and arm are not above the spinning chuck.
- Keep the floor around the machine clear of oil or grease to prevent anyone from slipping and falling into the machine.
- Turn the machine off before talking to anyone.

Turning operations

Turning is the removal of metal from the outer diameter of a rotating cylindrical workpiece. Turning is used to reduce the diameter of the workpiece, usually to a specified dimension, and to produce a smooth finish on the metal. Often the workpiece will be turned so that adjacent sections have different diameters.

Chucking the workpiece

We will be working with a piece of 1.5" diameter 45# medium carbon steel about 6" long. A workpiece which is relatively short compared to its diameter is stiff enough that we can safely turn it in the three-jaw chuck without supporting the free end of the work[4]. See Fig.3-6.

For longer workpieces, we would need to face and center-drill the free end and use a dead or live center in the tailstock to support it. Without such support, the force of the tool on the workpiece would cause it to bend away from the tool, producing a strangely shaped result. There is also the potential that the work could be forced loose in the chuck jaws and fly out as a dangerous projectile[5].

Fig.3-6 Chucking the workpiece

Insert the workpiece into the three-jaw chuck and tighten down the jaws until they just start to grip the workpiece. Rotate the workpiece to ensure that it is seated evenly. You want the workpiece to be as parallel as possible with the center line of the lathe spindle. Tighten the chuck using each of the three chuck key positions to ensure a tight and even grip. Remove the chuck key immediately after use.

Adjusting the tool bit

Choose a tool bit with a slightly rounded tip. This type of tool should produce a nice smooth finish. For more aggressive cutting, if you need to remove a lot of metal, you might choose a tool with a sharper tip. Make sure that the tool is tightly clamped in the tool holder.

Adjust the angle of the tool holder so that the tool is approximately perpendicular to the side of the workpiece. Because the front edge of the tool is ground at an angle, the left side of the tip should engage the work, but not the entire front edge of the tool[6].

Make sure the half-nut lever is disengaged. Move the carriage until the tip of the tool is near the free end of the workpiece, then advance the cross slide until the tip of the tool just touches the side of the work. Move the carriage to the right until the tip of the tool is just beyond the free end of the work.

Setting speed and feed

You must consider the rotational speed of the workpiece and the movement of the tool relative to the workpiece as discussed earlier. Based on Table 3-1 and the formula mentioned above, you can easily determine the rotational speed of the spindle. Basically, the softer the metal, the faster the cutting. Most cutting operations will be done at speeds of a few hundred RPM.

Turning with hand feed

As always, wear safety glasses and keep your face well away from the work since this operation will throw off hot chips and/or sharp spirals of metal.

Now advance the cross slide crank about 10 divisions or .010" (ten one-thousandths or one one-hundredth of an inch). Turn the carriage handwheel counterclockwise to slowly move the carriage towards the headstock. As the tool starts to cut into the metal, maintain a steady cranking motion to get a nice even cut. It's difficult to get a smooth and even cut turning by hand. See Fig.3-7.

Fig.3-7 Turning with hand feed

Continue advancing the tool towards the headstock until it is about 1/2" away from the chuck jaws. Obviously you want to be careful not to let the tool touch the chuck jaws!

Without moving the cross slide or compound, rotate the carriage handwheel clockwise to move the tool back towards the free end of the work. You will notice that the tool removes a small amount of metal on the return pass. Advance the cross slide another .010 and repeat this procedure until you have a good feel for it. Try advancing the cross slide by .020 on one pass. You will feel that it takes more force on the carriage handwheel when you take a deeper cut.

Turning with power feed

Turning with power feed will produce a much smoother and more even finish than is generally achievable by hand feeding. Power feed is also a lot more convenient than hand cranking when you are making multiple passes along a relatively long workpiece.

The power feed is engaged by the longitudinal & cross feed lever on the apron. Turn the motor on. The feed rod should be now rotating counterclockwise. With the tool positioned just beyond the end of the workpiece and advanced to make a cut of .010, engage the longitudinal & cross feed lever. The carriage should move slowly to the left under power from the feed rod. When the tool gets to within about 1/2" of the chuck, disengage the lever to stop the carriage motion. See Fig.3-8.

Fig.3-8 Turning with power feed

Now you can use the carriage handwheel to crank the carriage back to the starting point by hand. If you do so without first retracting the cutting tool, you will see that the tool cuts a shallow spiral groove along the workpiece. To avoid this, especially during finishing cuts, note the setting on the cross slide dial, then turn the cross feed crank a half turn or so counterclockwise to retract the tool. Now crank the carriage back to the starting point by hand, advance the cross slide back to the original dial setting plus an additional .010 and repeat the process. You should get a nice, shiny, smooth finish.

You normally will make one or more relatively deep (.010-.030) roughing cuts followed by one or more shallow (.001-.002) finishing cuts. Of course you have to plan these cuts so that the final finishing cut brings the workpiece to exactly the desired diameter.

When cutting under power, you must be very careful not to run the tool into the chuck. This seems to happen to everyone at one time or another, but it can shatter the tool and damage the chuck and will probably ruin the workpiece. So pay close attention and keep your hand ready on the lever.

Measuring the diameter

Most of time, a turning operation is used to reduce the workpiece to a specified diameter. It is important to recognize that, in a turning operation, each cutting pass removes twice the amount of metal indicated by the cross slide feed divisions[7]. This is because you are reducing the radius of the workpiece by the indicated amount, which reduces the diameter by twice that amount. Therefore, when advancing the cross slide by .010", the diameter is reduced by .020"[8].

The diameter of the workpiece is determined by a caliper or micrometer. Micrometers are more accurate, but less versatile. You will need a machinist's caliper capable of measuring down to .001". Vernier calipers do not have a dial and require you to interpolate on an engraved scale, while a dial caliper gives a direct easy-to-read and hard-to-misinterpret measurement. See Fig.3-9.

Fig.3-9 Measuring the diameter with a dial caliper

It should be self-evident that you should never attempt to measure the work while it is in motion. With the lathe stopped, bring the dial caliper up to the end and use the roller knob to close the caliper jaws down on the workpiece. It’s a good idea to take at least two separate measurements just to make sure you got it right.

Filing the Edge

Finally, you may want to use a file to make a nice beveled edge on the end of the workpiece. With the lathe running at fairly low speed, bring a smooth cut file up to the end of the workpiece at a 45 degree angle and apply a little pressure to the file. Be sure to hold the tang end of the file in your left hand so that your hand and arm are not above the spinning chuck.

TECHNICAL WORDS

groove	[gru:v]	*n. & v.*	槽；切槽
chamfer	['tʃæmfə]	*n. & v.*	斜面；斜切，倒角
knurl	[nə:l]	*n. & v.*	滚花
chip	[tʃip]	*n.*	切屑
install	[in'stɔ:l]	*v.*	安装
overhang	['əuvəhæŋ]	*n.*	伸出部分
spiral	['spairəl]	*n.& a.*	螺旋形（切屑）；螺旋形的
spin	[spin]	*v.*	旋转
file	[fail]	*n. & v.*	锉刀；锉削
grip	[grip]	*v. & n.*	紧握，抓紧
cylindrical	[si'lindrikəl]	*a.*	圆柱的，筒形的
circumference	[sə'kʌmfərəns]	*n.*	圆周，周长
perpendicular	[ˌpə:pən'dikjulə]	*a.*	垂直的
parallel	['pærəlel]	*a.*	平行的
crank	[kræŋk]	*n. & v.*	摇柄；摇手柄
counterclockwise	[ˌkauntə'klɔkwaiz]	*a. & adv.*	逆时针
clockwise	['klɔkwaiz]	*a. & adv.*	顺时针
retract	[ri'trækt]	*v.*	退回
shatter	['ʃætə]	*v.*	撞碎，毁坏
radius	['reidjəs]	*n.*	半径 radii ['reidiai](*pl.*)
diameter	[dai'æmitə]	*n.*	直径
caliper	['kælipə]	*n.*	卡钳
micrometer	[mai'krɔmitə]	*n.*	外径千分尺，测微计
interpolate	[in'tə:pəleit]	*v.*	插值；插入；插补
bevel	['bevəl]	*n. & v.*	（把……切成）斜边，斜面

PHRASES

depth of cut		切削深度
safety practice		安全习惯（惯例）
tool bit	（cutter bit）	刀具，刀头

rotational speed	旋转速度
rough(ing) cut	粗加工
finishing cut	精加工
cutting edge	切削刃
carbon steel tool	碳素工具钢刀具
chuck key	卡盘扳手
emergency stop	紧急停止（按钮）
tang end	（锉刀）柄脚
center-drill	打中心孔；中心钻
chuck jaw	卡盘卡爪
return pass	退刀
multiple passes	多次进刀
at one time or another	曾经
vernier caliper	游标卡尺
engraved scale	刻度尺
dial caliper	带表卡尺，带表盘卡尺

NOTES

1. The cutting speed of a tool bit is defined as the distance of the workpiece surface, measured at the circumference, that passes the tool bit in one minute. 刀具的切削速度定义为，一分钟内刀具经过工件表面接触的圆周距离。

2. If a large amount of stock is **to be removed**, it is advisable to take one or more roughing cuts and then take light finishing cuts at relatively high speeds. 如果毛坯切削量很大，建议进行一次以上的粗切削，然后以相对高速进行少量精切削。*to be removed 是被动语态的不定式形式，表示将要发生的动作。*

3. The depth of the cut for roughing is generally five to ten times deeper than the feed. The reason for this is that more of the cutting edge of the tool bit is in contact with the workpiece for the amount of metal being removed. 粗加工的切削深度通常是进给量的5~10倍，这么做的理由是刀具切削刃与工件接触更多，更利于去除金属量。*being removed 为现在分词作为后置定语。*

4. A workpiece **which is relatively short compared to its diameter** is stiff enough that we can safely turn it in the three-jaw chuck without supporting the free end of the work. 与其直径比起来相对短的工件刚性足够好，我们可以安全地车削它，装夹时只用三爪卡盘而不用支撑工件的自由端。*which is relatively short compared to its diameter 是定语从句，修饰 workpiece。*

5. Without such support, the force of the tool on the workpiece would cause it to bend away from the tool, producing a strangely shaped result. There is also the potential that the work could be forced loose in the chuck jaws and fly out as a dangerous projectile. 没有这样的支撑，刀具施加到工件上的力将使工件从刀具位置弯曲，结果制造出奇怪的形状。也有可能工件受力在卡爪

里松弛，像一个危险的抛射物一样飞出。

6. Because **the front edge of the tool** is ground at **an angle**, the left side of the tip should engage the work, but not the entire front edge of the tool. 由于刀具的前部刀刃（副切削刃）被磨出一个角度（副偏角），刀尖的左侧应该切入到工件，而不是整个前部刀刃（副切削刃）与工件接触。

7. It is important to recognize that, in a turning operation, each cutting pass removes twice the amount of metal indicated by the cross slide feed divisions. 在外圆车削过程中，每次进刀的金属去除量是中拖板进给刻度指示的两倍，认识到这一点很重要。*divisions 指拖板上刻度盘的刻度。*

8. This is because you are reducing the radius of the workpiece by the indicated amount, which reduces the diameter by twice that amount. Therefore, when advancing the cross slide by .010", the diameter is reduced by .020". 这是因为你减少的是根据刻度盘指示的工件半径，对直径来说，就减少了两倍的刻度指示量。因此，当中拖板进刀 0.010"时，直径就减少了 0.020"。*前一句话如果直译的话非常拗口，要根据上下文理解后整理成通顺的汉语。*

PRACTICE

Task 1　Translate the following words and phrases into English.

1．车外圆　　2．车内圆
3．车端面　　4．倒角
5．割断　　6．割槽
7．车锥面　　8．车螺纹
9．滚花　　10．打中心孔
11．切削速度　　12．进给量
13．切削深度　　14．自动进给
15．粗加工　　16．精加工
17．待切削工件　　18．所使用的车床
19．进刀(*v.*)　　20．退刀(*v.*)

Task 2　Fill in the brackets with words that have similar meaning to the underlined words, changing their forms if necessary.

1．(　　　　　　) Make sure that the tool is tightly clamped in the tool holder.

2．(　　　　　　) There is also the potential that the work could be forced to loosen in the chuck jaws and fly out as a dangerous projectile.

3．(　　　　　　) For a short workpiece, we can safely turn it in the three-jaw chuck without supporting the free end of the work.

4．(　　　　　　) When the tool gets to within about 1/4" of the chuck, disengage the lever to stop the carriage motion.

5．(　　　　　　) Turning with power feed will produce a much smoother and more even finish than is generally achievable by hand feeding.

6. (   ) The feed rod is driven by the spindle through a train of gears, and the ratio of its speed to that of the spindle can be varied by changing gears to produce various rates of feed.

7. (   ) Use the cross feed crank to back off the tool until it is beyond the diameter of the workpiece.

8. (   ) Never attempt to measure work while it is turning.

9. (   ) Avoid reaching over the spinning chuck.

10. (   ) Never attempt to break away metal spirals as they form at the cutting tool.

Task 3　Write the English name of each tool on the corresponding number.

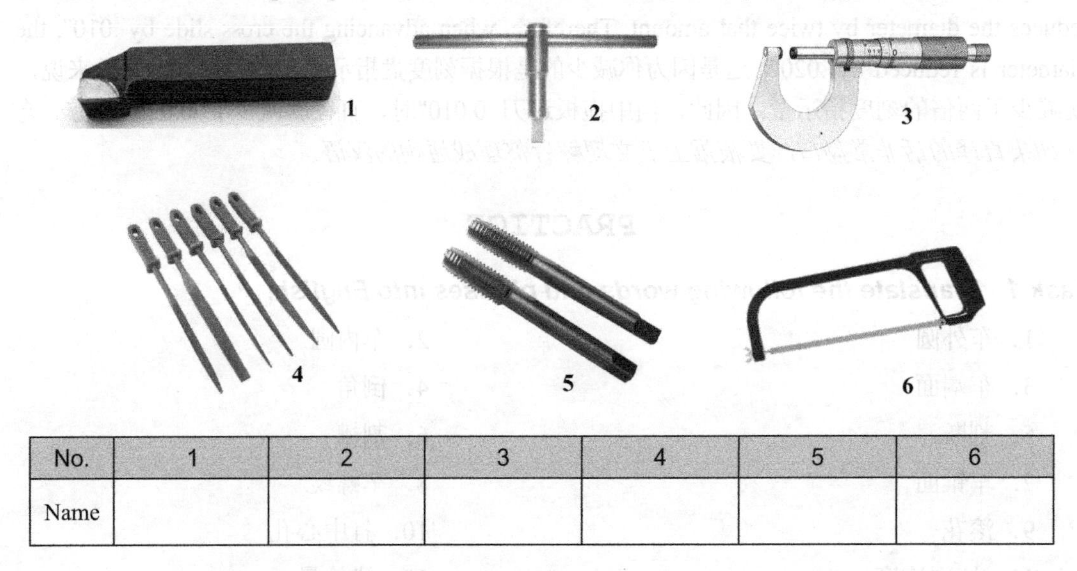

No.	1	2	3	4	5	6
Name						

Task 4　The turning operation may be broken down into several steps. Fill in the following table according to the text.

No.	Major steps	Key points		
		Success or failure	Safety	Techniques
1	Chucking the workpiece	Seated evenly; parallel with the spindle center line	Remove the chuck key immediately	Use the center in the tailstock for longer workpieces
2				
3				
4				
5				
6				

Task 5　Decide whether the following statements are true (T) or false (F).

1. (   ) Non-cylindrical shapes can not be faced on the lathe.

2. (　　　　) Light finishing cuts can be made at relatively high speeds.

3. (　　　　) To obtain uniform cutting speed, the lathe spindle must be revolved faster for workpieces of small diameters and slower for workpieces of large diameters.

4. (　　　　) Generally, the deeper the cut, the faster the speed, since a deep cut requires less power.

5. (　　　　) Basically, the softer the metal, the faster the cutting.

6. (　　　　) If the workpiece to be machined is short, we can safely turn it without the support of the center.

7. (　　　　) Micrometers are more accurate than calipers.

8. (　　　　) For the sake of safety, hold the tang end of the file in your left hand when filing on the lathe.

9. (　　　　) Turning with hand feed will produce more even finish than is generally achievable by power feed.

10. (　　　　) You may talk to someone while the lathe machine is working.

Task 6　Fill in the blanks with the appropriate form of the underlined words which can be of different form, or synonym, or acronym, etc.

*Example:　For finishing cuts, a light feed is necessary since a heavy feed produces a poor ___________. The correct answer should be **finish**.*

1. The turning machines ________ material from a rotating workpiece via the linear movements of various cutting tools. Turning is the removal of metal from the outer diameter of a rotating cylindrical workpiece.

2. Turning is used to reduce the workpiece to a specified diameter while facing is often used to bring the piece to a specified _________.

3. "V" and combination beds are used for precision and light duty work, while flat beds are used for ________ duty work.

4. The half-nuts are engaged only when the lathe is used to cut threads, at which time the feed mechanism must be ___________.

5. The apron contains the mechanism that controls the movement of the carriage for longitudinal feed and thread cutting. It controls the ________ movement of the cross slide.

6. These gear trains transmit power from the feed rod to move the carriage ______ the ways and to move the cross slide across the ways, thus providing powered longitudinal feed and cross feed.

7. Turning machines remove material from a rotating workpiece via the _______ movements of various cutting tools, such as tool bits and drill bits.

8. Turning with _________ feed will produce a much smoother and more even finish than is generally achievable by hand feeding.

9. Turn the carriage handwheel counterclockwise to slowly move the carriage towards the headstock. Rotate the carriage handwheel _________ to move the tool back towards the free end of the work.

10. Advance the cross slide by .020 on one pass. Turn the cross feed crank a half turn counterclockwise to ________ the tool.

Task 7 Choose the best answer.

1. When facing, which handwheel should be used?

(A) Carriage handwheel.

(B) Cross slide handwheel.

(C) Compound slide handwheel.

Your answer: ________

2. How much is the depth of the cut for roughing?

(A) Generally 2-3 times deeper than the feed.

(B) Generally 3-4 times deeper than the feed.

(C) Generally 5-10 times deeper than the feed.

Your answer: ________

3. The proper sequence of performing the turning operations for machinists is as the following:

① Turning with hand feed or power feed
② Adjusting the tool bit
③ Chucking the workpiece
④ Setting speed and feed
⑤ Measuring the diameter

(A) ③→②→④→①→⑤

(B) ①→②→④→③→⑤

(C) ①→②→③→④→⑤

Your answer:

4. Which cutter bits allow the highest cutting speed?

(A) Carbon steel.

(B) High-speed steel.

(C) Carbide tipped.

Your answer: ________

5. Which of the following statements is true?

(A) Power feed is a lot more convenient than hand cranking in some applications.

(B) Turning with hand feed will produce a much smoother and more even finish than is generally achievable by power feed.

(C) If carbon steel cutter bits are used, speeds may be 2 or 3 times as high as those given for high-speed steel cutter bits.

Your answer: __________

6. To obtain uniform cutting speed, how must the lathe spindle be revolved?

(A) The lathe spindle must be revolved slower for workpieces of small diameters and faster for workpieces of large diameters.

(B) The lathe spindle must be revolved faster for workpieces of small diameters and slower for workpieces of large diameters.

(C) The rotational speed of the lathe spindle has no relationship with the cutting speed.

Your answer: __________

7. Normally, which material being worked on may select the highest lathe cutting speeds for straight turning?

(A) Aluminum. (B) Cast iron. (C) Stainless steel.

Your answer: __________

8. Normally, which material being worked on may select the lowest lathe cutting speeds for straight turning?

(A) Aluminum. (B) Cast iron. (C) Stainless steel.

Your answer: __________

9. When advancing the cross slide by .020", the diameter is reduced by __________.

(A) .040" (B) .010" (C) .020"

Your answer: __________

10. To slowly move the carriage towards the headstock, you should:

(A) Turn the carriage handwheel clockwise

(B) Turn the carriage handwheel counterclockwise

(C) Turn the cross slide crank counterclockwise

Your answer: __________

11. The cross slide of a lathe is usually used to control:

(A) Feeds (B) Speeds (C) Depth of cut

Your answer: __________

12. When you rotate the carriage handwheel clockwise of the lathe,

(A) The tool is moved towards the free end of the work

(B) The tool is moved towards the headstock

(C) The tool is moved towards the center of the work

Your answer: __________

13. In machining operations, there is one sequence of events that one must always follow:

(A) ACCURACY FIRST, SAFETY SECOND, AND SPEED LAST

(B) SAFETY FIRST, ACCURACY SECOND, AND SPEED LAST

(C) SAFETY FIRST, SPEED SECOND, AND ACCURACY LAST

Your answer: __________

14. Of the following cutting conditions, which one has the greatest effect on tool wear?

(A) Cutting speed. (B) Depth of cut. (C) Feed.

Your answer: __________

15. Feed is specified in:

(A) mm/r (B) mm/min (C) r/min

Your answer: __________

16. For a workpiece of 100mm diameter, if you want to reduce the diameter to 94mm on one pass, then the depth of cut will be:

(A) 94mm (B) 6mm (C) 3mm

Your answer: __________

17. We will be working with a piece of 1.5" diameter 45# steel. If the selected cutting speed is 24m/min, then the spindle speed of the lathe should be set to:

(A) 200r/min (B) 250r/min (C) 300r/min

Your answer: __________

18. For most materials, the feed for rough cuts should be:

(A) 0.010 to 0.020 inch per revolution

(B) 0.010 to 0.020 millimeter per revolution

(C) 0.10 to 0.20 inch per revolution

Your answer: __________

19. The thread of the lead screw of the cross slide is 5mm, and the cross slide dial has 100 divisions. If advancing the cross slide crank 10 divisions, then how much is the diameter of the workpiece reduced by?

(A) 1mm (B) 2mm (C) 0.5mm

Your answer: __________

20. Which of the following statements is not true?

(A) Advancing the cross slide by .010 on one pass makes a deeper cut than advancing the cross slide by .020 on one pass.

(B) It takes more force on the carriage handwheel when you take a deeper cut.

(C) You normally will make one or more relatively deep (.010-.030) roughing cuts followed by one or more shallow (.001-.002) finishing cuts.

Your answer: __________

Task 8 You are required to select one word for each blank from a list of choices given in a word bank following the passage. Each choice in the bank is identified by a letter. You may not use any of the words in the bank more than once. Read aloud the passage when completed.

Use the compound handwheel to advance the tip of the tool until it just touches the end of the (1)________. Use the cross feed (2)________ to back off the tool until it is beyond the diameter of the workpiece. Turn the lathe on and adjust the speed to a few hundred (3)________. Now slowly advance the (4)________ handwheel to move the tool towards the workpiece. When the tool touches the workpiece it should start to remove (5)________ from the end. See Fig.3-10. Continue advancing the (6)________ until it reaches the (7)________ of the workpiece and then crank the tool back in the (8)________ direction (towards you) until it is back past the (9)________ of the workpiece.

A. cross feed	F. center
B. RPM	G. opposite
C. speed	H. edge
D. tool	I. metal
E. workpiece	J. crank

Fig.3-10 Facing cut

Task 9 Problem. The thread of the lead screw of the cross slide is 5mm, and the cross slide dial has 100 divisions.

1. If advancing the cross slide crank one division, then how much is the tool advanced?

----

2. If advancing the cross slide crank 20 divisions, then how much is the diameter of the workpiece reduced by?

----

3. If the workpiece diameter is to be reduced from 70mm to 65mm on one pass, then how many divisions is the cross slide crank supposed to be advanced?

Part B Listening

Task 1 Listen to the five statements twice and write them down.

1. --
2. --
3. --
4. --
5. --

Task 2 The following video clip is about lathe safety. Watch it first, then listen to it twice and write down the safety tips according to what you see and hear.

1. --
2. --
3. --

4. ....

5. ....

Task 3 The following video clip is about work holding on the lathe. Watch it first, then listen to it twice and fill in the blanks with what you hear and see.

A three-jaw (1)__________ moves its jaws in or out in exact unison（一致）. For larger parts the jaws on the three-jaw chuck can be (2)__________. A four-jaw chuck has independently adjustable (3)__________. It can be used to (4)__________ parts that are not (5)__________. The (6)__________ part is far off center. We (7)__________ each jaw accordingly and verify that it is (8)__________.

Task 4 The following video clip is about speeds and feeds setting. Watch it first, then listen to it twice and fill in the blanks with what you hear and see.

Nearly every gear head engine lathe uses the method similar to this machine for setting (1)__________. First you have to locate the spindle speed readout（读数） on the machine. Now select an (2)__________ from those available. Here we select (3)__________ RPM which requires these two (4)__________ to be in the correct position. We now change speed to (5)__________ RPM.

Feed rates can be set by (6)__________ the settings panel（面板） on the lathe. Here we pick (7)__________ per revolution. The four (8)__________ and one (9)__________ must be set to (10)__________. See Fig.3-11.

Fig.3-11 Settings panel

Part C Speaking

Task 1 Watch the slides and give the English name or description for each of the slides. Take notes.

1. .... 3. ....

2. .... 4. ....

5. .... 8. ....

6. .... 9. ....

7. .... 10. ....

Task 2 Read aloud the following numbers, symbols, and expressions. Take notes.

No.	Number/symbol	Pronunciation	No.	Number/symbol	Pronunciation
1	0.1		15	$X+$	
2	0.01		16	$Z-$	
3	0.001		17	1/4"	
4	2.35		18	1.5"	
5	1/4		19	0.01"	
6	1/10		20	0.001"	
7	1/100		21	0.005"	
8	+		22	1:2	
9	−		23	90°	
10	×		24	sin30°	
11	÷		25	1000RPM	
12	=		26	80m/min	
13	±		27	V_C	
14	>		28	$S=\frac{1000\cdot V_C}{\pi\cdot D}$	

Task 3 The following video is about lathe operations. Watch it twice and write down the operations involved according to what you see. Complete the following table. After that, tell the class what's about the video just as a narrator.

No.	Operations
1	
2	
3	
4	
5	
6	

Task 4 Watch the video clip about work holding with the loudspeaker mute and tell the class how to chuck the workpiece on the lathe just as a narrator.

Task 5 Look at the following safety symbols and decide what each symbol stands for. You may use such sentence pattern as "this symbol means/indicates/ warns…,etc."

1 2 3 4 5 6 7 8 9 10 11 12

1. ....
2. ....
3. ....
4. ....
5. ....
6. ....
7. ....
8. ....
9. ....
10. ....
11. ....
12. ....

Task 6 Work in pairs. Explain in English the following technical words and phrases. Select some of them to make a situational dialogue with the help of the pictures on page 41. Take notes. Role-play the dialogue before the class. You may search the information on the Internet.

cross slide, turning, engine lathe, facing, threading, boring, caliper

Task 7 Shop floor practice. Work in groups. Discuss the working plan for cylindrical turning. Fill out the following table. Make the plan as detailed as possible.

Trainee name(s): ______________________________

工序号 Item	工序内容 Operations	夹紧工具 Clamping tools	切削刀具 Cutting tools	刀具材料 Tool material	切削速度 Cutting speed	进给量 Feed	切削深度 Depth of cut	测量工具 Measuring tools
1								
2								
3								
4								
5								
6								
7								
8								
9								
10								

Task 8 Work in pairs. Create a technical situation using job instruction and role-play the dialogue before the class. One acts as the master, another as the apprentice.

WORKING SITUATION

How to pass along our skills to others? You need give **job instruction**.

【引导文】工作指导的四阶段法 Job Instruction

1. 在现场工作培训中运用四阶段法。

Steps	Actions
第一阶段：学习准备 Prepare the learner	使学习者轻松愉快 Put the learner at ease 告诉他将做何种工作 State the job 了解他对这项工作的认识程度 Find out what they already know 激发他学习这项工作的兴趣 Get the team member interested in learning 使他进入正确的学习位置 Put the learner in the correct position
第二阶段：传授工作 Present the operation	将主要步骤，一步一步地讲给他听，做给他看 Instructor does the job and describes the operation one major step at a time 明确强调要点 Instructor does the job, states the major steps, and stresses each key point 说明要点的理由 Instructor does the job, states the major steps, key points, and explains the reasons
第三阶段：尝试练习 Tryout performance	让他试做，纠正错误 Learner does the job silently. Instructor corrects errors as needed. 让他边做边说出主要步骤 Learner does the job, explains each major step 让他边做边说出要点 Learner does the job, explains the major steps, & key points behind each step 让他说明要点的理由，并确认他完全掌握 Learner does the job, explains the major steps, key points, & reasons why
第四阶段：检查成效 Follow up	安排他开始具体工作 Put learner on own 鼓励他提出问题 Encourage questions by the learner 指定可以帮助他的人 Designate who to go to for help 经常检查 Check back frequently at first 逐渐减少指导的次数 Taper off and provide coaching as needed

2．要有效地开展工作指导，需要制作作业分解表 Job Breakdown Sheet，以下是一个关于万用表使用的作业分解表样例。

作业分解表 Job Breakdown Sheet

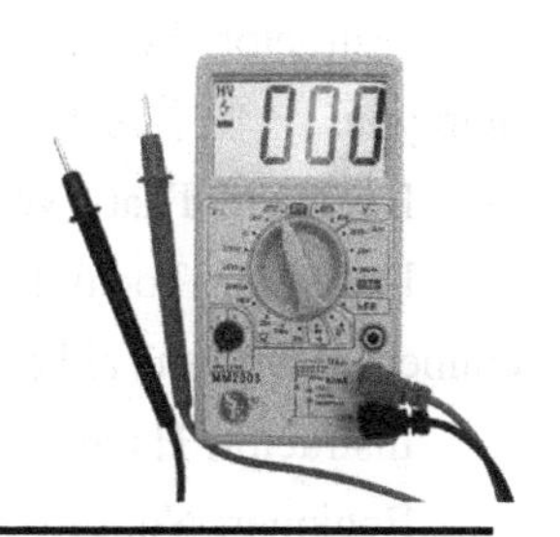

作　　业 Operation　用万用表判断导线连接的正确性

部　　件 Parts　电气线路

工具及材料 Tools & materials　万用表+双股导线

主要步骤（Major steps） 能促使工作顺利完成的主要作业顺序	要点（Key points） 根据步骤找出如下要点 （1）左右工作能否完成的作业内容——成败 （2）危险——作业员有可能受伤的关键点——安全 （3）具备能使工作顺利完成的技术——易做
1. 打开万用表电源（开） Turn on the multimeter	检查测量表棒线是否插好——成败 Check if probes are properly connected
2. 选择电阻挡（选） Set the multimeter to Ohms with selector	看标记“Ω”处——成败 Find the label Ω
3. 将两根表棒搭在一起（调） Get the test leads contact with each other	看读数是否由 1. 变为 0.000——成败 Check if the meter readout becomes 0.000
4. 用两表棒分别接触导线的两端（量） Place one probe on each end of a circuit	（1）测量电阻时，必须切断电源——成败 Make sure the item being repaired is turned off
	（2）表棒的金属部分与导线的铜丝部分必须可靠接触——成败 Make sure the metal parts of the two ends contact
	（3）测量时，没电当有电操作——安全 Always keep electrical safety in mind even when the item is turned off
	（4）仅用一只手抓住两根表棒——易做 Grasp the two test leads with only one hand
5. 看万用表显示的读数（看） Observe the indicator	如果显示 0.000，则导线连接良好；如果仍为 1. ，则导线连接断线——成败 测量原理：导线相通时，电阻接近 0；导线不通时，电阻为∞ When the circuit is electrically connected, the readout will be nearly 0Ω; if there is a break in the circuit, the readout will display an infinite amount of resistance
6. 关闭万用表电源（关） Turn off the meter	省电——安全 Set the multimeter to other than Ohms with selector.

3. 工作指导实例（对话）。

Instructor: Nice to meet you. My name is Tang Caiping. I'm responsible for CNC maintenance and application in this company. Welcome to our team.

Beginner: Nice to meet you, too.

Instructor: You are one of our members now. So please feel at home. Every one will be glad to help you. （使学习者轻松愉快）

Beginner: That's very kind of you.

Instructor: Today I will instruct you how to judge if the wiring in electrical circuits is properly connected with the aid of a multimeter. （告诉他将做何种工作）

Instructor: Have you ever used the multimeter?（了解他对这项工作的认识程度）

Beginner: No.

Instructor: The multimeter is the electrician's best friend. A versatile tool, the multimeter is a must-have if you plan on doing any of your own electrical work in your home. You may want to do some repairing when your home appliances do not work. （激发他学习这项工作的兴趣）

Beginner: Oh, That's great. I'm eager to know how.

Instructor: Now please stand on my right. We are going to take a closer look at the multimeter and explain how you can get the most use out of this essential device. （使他进入正确的学习位置）

Instructor: Now let's become familiar with the parts of a multimeter. Test leads. There should be 2 test leads or probes. Generally, one is black and the other red. And this is the LCD display to show the readout.

Instructor: The first step is to turn the meter on. （将主要步骤一步一步地讲给他听，做给他看）

Instructor: The second step is to set the multimeter to Ohms with the selector.

Instructor: The third step is to get the test leads in contact with each other and observe the meter readout. If it becomes 0.000, the meter is working.

Instructor: The fourth step is to place one probe on each end of a circuit.

Instructor: The fifth step is to observe the indicator. If the meter reads **1.**, then there is a break in the circuit. If you get a numerical reading along with an audible tone, then the circuit is electrically connected.

Instructor: The sixth step is to turn the meter off.

Instructor: So do you remember the six steps?

Instructor: I will now explain the key points. Before we start our work, let's check if the probes are properly connected. The negative probe is black and the positive will be red. The black probe ALWAYS gets plugged into the terminal that is labeled "COM" or common. （明确强调要点）

Instructor: If the test leads are not in contact with anything, the readout will be **1.**. This represents an infinite amount of resistance, or an "open circuit".

Instructor: The item being repaired should be turned off, or you will damage the multimeter. And always be mindful of electrical safety even when the item is turned off.

Instructor: When not in use, store probes so they will not be damaged and do not forget to set the multimeter to other than Ohms with selector so the battery will not run down.

Instructor: Now it's your turn to have a try. As you test, please talk us through the procedures.

Beginner: First turn on the meter. Select the Ohms setting. Next get the test leads in contact with each other and observe the meter readout. Now it becomes 0.000, so the meter is in good state. Then place one probe on each end of the circuit. Observe the indicator. Now the meter gives a numerical reading, then the circuit is connected. Finally, turn the meter off when not in use and keep the meter in other range than Ohms for saving the battery. （让他边做边说出主要步骤）

Instructor: OK. Please do it once more and tell the key points and why as well. （让他边做边说出要点）

Beginner: Before turning on the meter, make sure that the probes are properly connected. The item being repaired should be turned off, or the multimeter may be destroyed. If the test leads are not in contact with anything, the readout will be 1.. This represents an infinite amount of resistance, or an "open circuit".

Instructor: Very good. You did a very good job. Have you still got any questions? （鼓励他提出问题）

Beginner: How can I confirm that the probes are connected properly?

Instructor: Just check the readout. If the probes are not properly connected, then no readout changes take place even if you get the test leads in contact with each other.

Beginner: Oh, I got it. Thank you very much.

Instructor: Then you may get to work. Mr. Ma is an experienced master here. If you have problems, don't hesitate to ask him for help. I will come often and I hope you enjoy the work here. （安排他开始具体工作，指定可以帮助他的人）

Beginner: Thank you. I will take it seriously.

4. 常用的工作指导用语

Have you got it? 明白了吗？

I've got it. 我懂了。

I see．我明白了。

I can't follow you． 我不懂你说的。

I have no idea． 我没有头绪。

Let me see．让我想想。

I beg your pardon? 请您再说一遍（我没有听清）。

选择一项简单的专业操作（如车外圆），对新手进行培训（时间控制在8~10分钟以内）。

作业分解表 Job Breakdown Sheet

作　　业 Operation ________________________________

部　　件 Parts ________________________________

工具及材料 Tools & materials ________________________________

主要步骤（Major steps） 能促使工作顺利完成的主要作业顺序	要点（Key points） 根据步骤找出如下要点： （1）左右工作能否完成的作业内容——成败 （2）危险——作业员有可能受伤的关键点——安全 （3）具备能使工作顺利完成的技术——易做
1.	
2.	
3.	
4.	
5.	
6.	
7.	
8.	
9.	

Role-play the job instruction.

Instructor:

Beginner:

Instructor:

Beginner:

Part D Grammar and Translation

被动语态的译法

专业英语中大量使用被动语态，这是因为文章需要客观地叙述事理，而不是强调动作的主体。

The clutch controls the direction of spindle rotation. 离合器控制主轴旋转方向。

The direction of spindle rotation **is controlled** by the clutch. 主轴旋转方向由离合器控制。

在第一句中，动词的主语“the clutch”实施动词表示的动作，这个动词称为处于“主动语态”；在第二句中，动词的动作被施加到主语上，该动词称为处于“被动语态”。

只有及物动词（其后可跟一个宾语）可用于被动语态。

主动形式的**宾语**，通常是被动形式的**主语**。

各种时态的被动语态如下。

一般现在时的被动语态: Computer **is used** in business, governments and institutions. 计算机用于商业、政府和公共机构。

现在进行时的被动语态: Computers **are being used** in machine tool control. 计算机正用于机床的控制。

过去进行时的被动语态: The hard-wired NC **was being used** widely in 1960s. 20 世纪 60 年代时，硬线数控应用很广。

不定式的被动语态: If a large amount of stock **is to be removed**, it is advisable to take one or more roughing cuts and then take light finishing cuts at relatively high speeds. *见 NOTES 2.*

带情态动词的被动语态: Finishing cuts are generally very light; therefore, the cutting speed **can be increased** since the chip is thin. 精切削一般量很小，由于切屑很薄，因此可以提高切削速度。

短语动词的被动语态: The documents attached to the machine should **be taken good care of.** 必须妥善保管这些随机资料。

把英语被动句译成汉语时，一般可以采用下列处理方法。

1. 译成汉语被动句

用“被”、“由”、“受”、“靠”、“给”、“遭”等汉语中表达被动概念的介词引导出施动者。

The power feed is engaged by the longitudinal& cross feed lever on the apron. 自动进给由溜板箱上的纵向/横向进给手柄啮合。

The diameter of the workpiece is determined by a caliper or micrometer. 工件直径由卡尺或千分尺确定。

2. 译成汉语无主语句

Good safety practices should be followed to ensure safe machining. 必须遵守良好的安全规程以保证安全加工。

3．增译“人们”、“有人”、“操作员”等主语

It is recommended that you never remove your hand from the chuck key when it is in the chuck. 我们建议只要卡盘扳手在卡盘上，手就不要离开扳手。

4．由 by 或 in 引导的状语往往可以转换为汉语的主语

M codes are commonly used by the machine tool builder to give the user programmable ON/OFF switches for machine functions.（Ref. LS 4）机床制造商通常用 M 代码给用户提供可编程的机床开/关功能。

Information regarding the machine's construction is usually published right in the machine tool builder's manual. （Ref. LS 4）机床制造厂的说明书通常发布机床结构方面的信息。

5．将英语句中的一个适当成分译成汉语中的主语

有时英文里的被动含义在中文里不一定需要表示出来。

The cutting tool is moved a definite distance along the work for each revolution of the spindle. 主轴每转一转，刀具沿工件移动固定的距离。

The apron is attached to the front of the carriage. 溜板箱连接在大拖板的前方。

Task　Translate the following sentences in passive voice into Chinese.

1. The carriage is then moved by the thread of the lead screw instead of by the gears of the apron feed mechanisms.

----

2. The half-nuts are engaged only when the lathe is used to cut threads, at which time the feed mechanism must be disengaged.

----

3. The cutting tool is moved a definite distance along the work for each revolution of the spindle.

----

4. Any number of threads can be cut by merely changing the gears in the connecting gear train to obtain the desired ratio of the spindle and the lead screw speeds.

----

5. The temperature of the workpiece is a key concern because most metals expand when heated.

----

6. The workpiece diameter is reduced by twice the depth of cut in each complete traverse of the tool bit.

----

7. Always double check to make sure your work is securely clamped in the chuck or between centers before starting the lathe.

----

8. Often the workpiece will be turned so that adjacent sections have different diameters.

Part E Supplementary Reading

Facing operations

Facing is the process of removing metal from the end of a workpiece to produce a flat surface. Most often, the workpiece is cylindrical, but using a four-jaw chuck you can face rectangular or odd-shaped（形状奇怪的） work to form cubes（立方形的东西） and other non-cylindrical shapes. Facing is often used to bring the piece to a specified length.

Chucking the workpiece

Clamp the workpiece tightly in the three-jaw chuck. To get the work properly centered, close the jaws until they just touch the surface of the work, then rotate the workpiece by hand in the jaws to seat it; then tighten the jaws. It's good practice to tighten the jaws from all three chuck key positions to ensure even gripping by the jaws. When a lathe cutting tool removes metal it applies considerable lateral（侧面的） force to the workpiece. To safely perform a facing operation the end of the workpiece must be positioned close to the jaws of the chuck. The workpiece should not extend more than 2-3 times its diameter from the chuck jaws unless a steady rest（中心架） is used to support the free end. See Fig.3-12.

Preparing for the facing cut

Choose a cutting tool with a slightly rounded tip. A tool with a sharp pointed tip will cut little grooves across the face of the work and prevent you from getting a nice smooth surface. Clamp the cutting tool in the tool post and turn the tool post so that the tip of the cutting tool will meet the end of the workpiece at a slight angle. It is important that the tip of the cutting tool be right at the horizontal center line of the workpiece; if it is too high or too low, you will be left with a little bump（隆起物） at the center of the face.

Clamp the tool post in place and advance the carriage until the tool is about even with the end of the workpiece.

Set the lathe to its lowest speed and turn it on.

Beginning the facing cut

Use the compound handwheel to advance the tip of the tool until it just touches the end of the workpiece. Use the cross feed crank to back off the tool until it is beyond the diameter of the workpiece. Turn the lathe on and adjust the speed to a few hundred RPM. Now slowly advance the cross feed handwheel to move the tool towards the workpiece. When the tool touches the workpiece it should start to remove metal from the end. See Fig.3-13. Continue advancing the tool until it reaches the center of the workpiece and then crank the tool back in the opposite direction (towards you) until it is back past the edge of the workpiece.

The roughing cut

Use the compound crank to advance the tool towards the chuck about .010" (ten one-thousandths of an inch, or one one-hundredth of an inch). If the compound is set at a 90 degrees to the cross slide then each division you turn the crank will advance the tool .001" (one one-thousandth of an inch) toward the chuck.

Fig.3-12 Chucking the workpiece to be faced

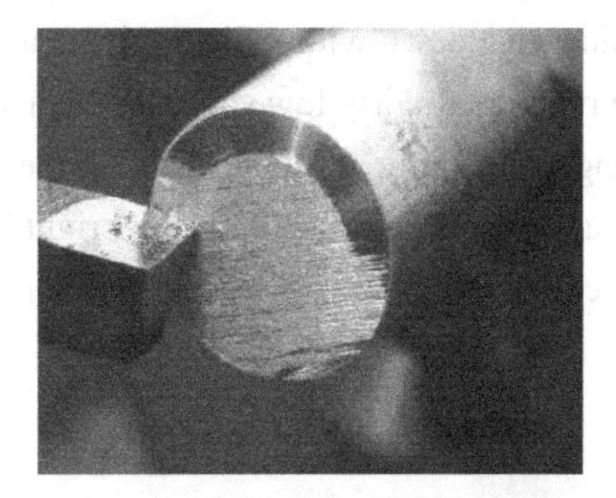

Fig.3-13 Beginning the facing cut

If the compound is set at some other angle, say 30 degrees, to the cross slide, then it will advance the tool less than .001" for each division. The exact amount is determined by the trigonometric sine（三角正弦） of the angle. Since the sine of 30 degrees is .5 the tool would advance .0005" (five ten-thousandths or 1/2 of one one-thousandth of an inch) for each division in this example. See Fig.3-14.

The finishing cut

Depending on how rough the end of the workpiece was to begin with and how large the diameter is, you may need to make 3 or more passes to get a nice smooth finish across the face. These initial passes are called roughing passes and remove a relatively large amount of metal.

When you get the face pretty smooth you can make a final finishing cut to remove just .001 to .003" of metal and get a nice smooth surface. The finishing cut can also be made at higher RPM (say 1500 RPM) to get a smoother finish. See Fig.3-15.

Fig.3-14 The first pass of a facing operation

Fig.3-15 A finishing cut in progress

Fig.3-16 shows what happens if the tip of your cutting tool is below the center line of the lathe - a little nub is left at the center of the workpiece. The same thing happens if the tool is too high but the nub will have more of a cone（圆锥体） shape in that case. If the tool is too low, place a suitable thickness of shim（薄垫片）stock underneath the tool in the tool holder. If it's too high, grind the top down a few thous.

Filing the edge

Facing operations leave a rather sharp edge on the end of the workpiece. It's a good idea to smooth this edge down with a file to give it a nice chamfer and to avoid cutting yourself on it. With the lathe running at fairly low speed, bring a smooth cut file up to the end of the workpiece at a 45 degree angle and apply a little pressure to the file. See Fig.3-17. Be sure to hold the tang end of the file in your *left* hand and the tip in your right hand to avoid having your left hand reaching over the spinning chuck.

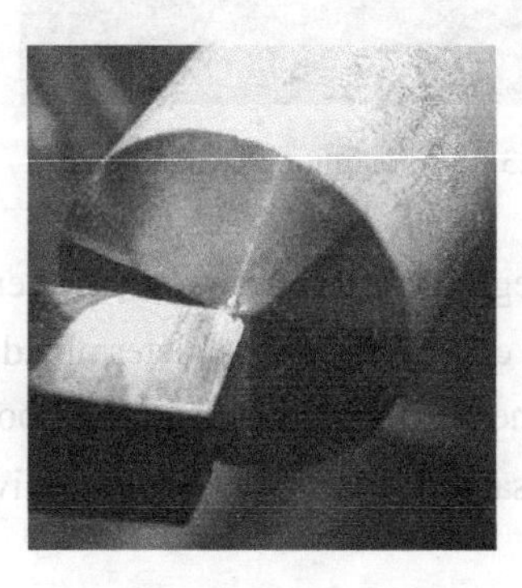

Fig.3-16 A little nub is left at the center of the workpiece

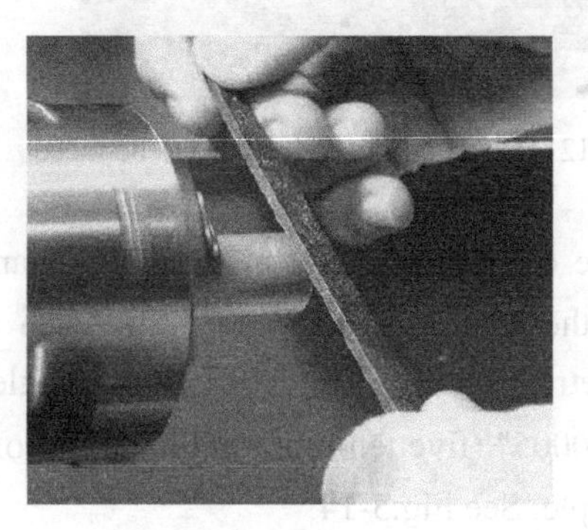

Fig.3-17 Filing the edge

Learning Situation 4

Know your CNC machine

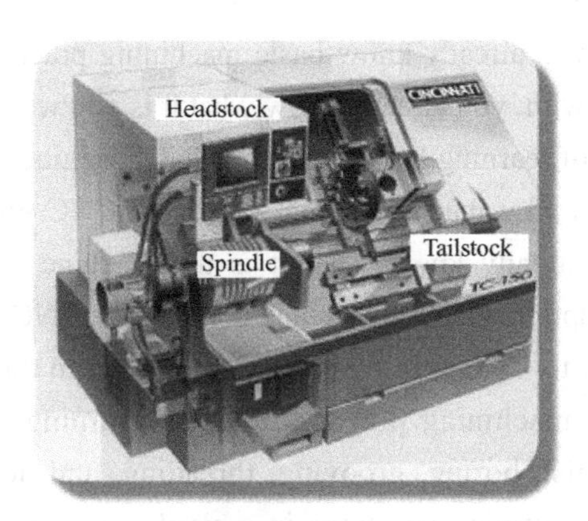

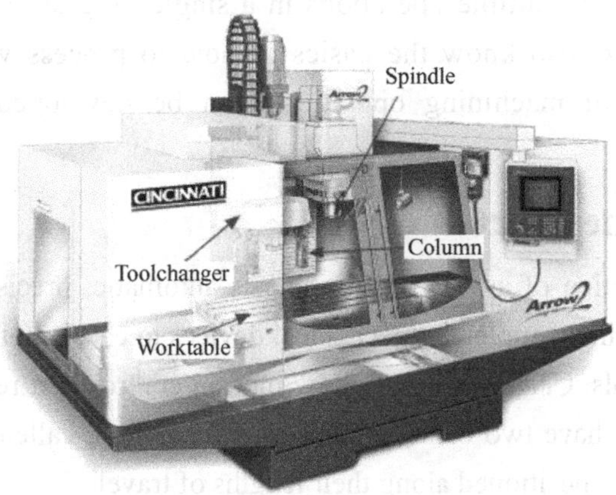

Focus of the situation

The key to success with any CNC machine is basic machining practice. Familiarity with the machine's capacity and construction, axes, accessories, and programmable functions should be acquired. [要想用好一台数控机床，必须具备基本的机械加工实践经验，同时对所用机床的规格、结构、坐标轴、附件、辅助功能有清楚的了解。]

Field work

Find a CNC machine and the machine tool builder's manual attached, getting familiarized with its construction, specifications and features.

Part A Reading

A CNC user must understand the basic operating principle and the makeup of the CNC machine tool being utilized.

Basic machining practice - the key to success with any CNC machine

Many forms of CNC machines are designed to enhance or replace what is currently being done with more conventional machines. The first goal of any CNC beginner should be to understand the basic machining practice that goes into using the CNC machine tool. The more the beginning CNC user knows about basic machining practice, the easier it will be to adapt to CNC[1].

Think of it this way. If you already know basic machining practice as it relates to the CNC machine you will be working with, you already know what it is you want the machine to do[2]. It will be a relatively simple matter of learning how to tell the CNC machine what it is you want it to do (learning to program). This is why machinists make the best CNC programmers, operators, and setup personnel. Machinists already know what it is the machine will be doing. It will be a relatively simple matter of adapting what they already know to the CNC machine.

For example, a beginner to CNC turning centers (as shown in the upper picture on page 73) should understand the basic machining practice related to turning operations like rough and finish turning, rough and finish boring, grooving, threading, and necking. Since this form of CNC machine can perform multiple operations in a single program (as many CNC machines can), the beginner should also know the basics of how to process workpieces machined by turning so a sequence of machining operations can be developed for workpieces to be machined[3].

Motion control - the heart of CNC

The most basic function of any CNC machine is automatic, precise, and consistent motion control. Rather than applying completely mechanical devices to cause motion as is required on most conventional machine tools, CNC machines allow motion control in a revolutionary manner[4]. All forms of CNC equipment have two or more directions of motion, called axes. These axes can be precisely and automatically positioned along their lengths of travel.

Instead of causing motion by turning cranks and handwheels as is required on conventional machine tools like engine lathes, CNC machines allow motions to be commanded through programmed commands. Generally speaking, the motion type (rapid, linear, and circular), the axes to move, the amount of motion and the motion rate (feed rate) are programmable with almost all CNC machine tools.

A CNC command executed within the control tells the drive motor to rotate a precise number of times. The rotation of the drive motor in turn rotates the ball screw. And the ball screw drives the linear axis (slide). A feedback device (linear scale) on the slide allows the control to confirm that the commanded number of rotations has taken place[5]. Refer to Fig.4-1.

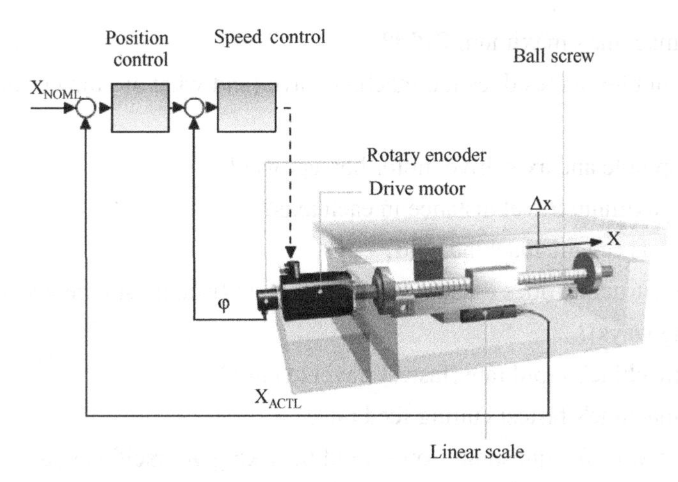

Fig.4-1 Principle of motion control

Though a rather crude analogy, the same basic linear motion can be found on a common table vise. As you rotate the vise crank, you rotate a lead screw that, in turn, drives the movable jaw on the vise. By comparison, a linear axis on a CNC machine tool is extremely precise. The number of revolutions of the axis drive motor precisely controls the amount of linear motion along the axis.

Learning about a new CNC machine - the key points

From a programmer's standpoint, as you begin to learn about any new CNC machine, you should concentrate on four basic areas. First, you should understand the machine's most basic components. Second, you should become comfortable with your machine's directions of motion (axes). Third, you should become familiar with any accessories equipped with the machine. And fourth, you should find out what programmable functions are included with the machine and learn how they are programmed.

Machine components

Fig.4-2 Slant bed style

The two most common CNC machines are the turning center and the machining center. A turning center is used to machine cylindrical parts, and a machining center is used to machine flat or angled surfaces. Figures on page 73 illustrate both of these machines. For a universal style slant bed turning center (as shown in Fig.4-2), for example, the programmer should know the most basic machine components, including the bed, way system, headstock & spindle, turret construction, tailstock, work holding device and CNC control. For the machining center, it includes the bed, saddle, column, worktable, servo motors, ball screws, spindle, tool changer, and CNC control. Information regarding the machine's construction including assembly drawings is usually published right in the machine tool builder's manual. As you read the machine tool builder's manual, here are some of the machine capacity and construction questions to which you should find answers.

- What is the machine's maximum RPM?
- How many spindle ranges does the machine have (and what are the cut-off points for each range?
- What is the spindle and axis drive motor horsepower?
- What is the maximum travel distance in each axis?
- How many tools can the machine hold?
- What way construction does the machine incorporate (usually square ways, dovetail, and/or linear bearing ways)?
- What is the machine's rapid rate (fastest traverse rate)?
- What is the machine's fastest cutting feed rate?

These are but a few of the questions you should be asking yourself as you begin working with any new CNC machine. Truly, the more you know about your machine's capacity and construction, the easier it will be to get comfortable with the machine.

Directions of motion (axes)

The CNC programmer must know the programmable motion directions (axes) available for the CNC machine tool. The axis names will vary from one machine tool type to the next. They are always referred to with a letter address. Common axis names are X, Y, and Z for linear axes and A, B, and C for rotary axes. These are related to the coordinate system. See Fig.4-3. However, the beginning programmer should confirm these axis designations and directions (plus and minus) in the machine tool builder's manual since not all machine tool builders conform to the axis names we show.

Fig.4-3 Coordinate system

The reference point for each axis

Most CNC machines utilize a very accurate position along each axis as a starting point or reference point for the axis. Some control manufacturers call this position the zero return position. Others call it the grid zero position. Yet others call it the home position. Regardless of what it is called, the reference position is required by many controls to give the control an accurate point of reference. CNC controls that utilize a reference point for each axis require that the machine be manually sent to its reference point in each axis as part of the power up procedure[6]. Once this is completed, the control will be in sync with the machine's position.

Accessories to the machine

Examples of CNC accessories include probing systems, tool length measuring devices, post process gauging systems, automatic pallet changers, adaptive control systems, bar feeders for turning centers, live tooling and C axis for turning centers, and automation systems. Truly, the list of potential accessory devices goes on and on.

Programmable functions

The programmer must also know what functions of the CNC machine are programmable. With low cost CNC equipment, often times many machine functions must be manually activated. With some CNC milling machines, for example, about the only programmable function is axis motion. Just about everything else may have to be activated by the operator. With this type of machine, the spindle speed and direction, coolant and tool changes may have to be activated manually by the operator.

With full-blown CNC equipment, on the other hand, almost everything is programmable and the operator may only be required to load and remove workpieces. Once the cycle is activated, the operator may be freed to do other company functions.

Reference the machine tool builder's manual to find out what functions of your machine are programmable. To give you some examples of how many programmable functions are handled, here is a list of a few of the most common programmable functions along with their related programming words.

Spindle control

An “S” word is used to specify the spindle speed (in RPM for machining centers). An M03 is used to turn the spindle on in a clockwise (forward) manner. M04 turns the spindle on in a counter clockwise (reverse) manner. M05 turns the spindle off. Note that turning centers also have a feature called constant surface speed which allows spindle speed to also be specified in surface feet per minute (or meters per minute).

Automatic tool changer (machining center)

A “T” word is used to tell the machine which tool station is to be placed in the spindle. On most machines, an M06 tells the machine to actually make the tool change. A four digit “T” word is used to command tool changes on most turning centers. The first two digits of the T word specify the turret station number and the second two digits specify the offset number to be used with the tool. T0101, for example, specifies tool station number one with offset number one.

Coolant control

M08 is used to turn on the flood coolant. If available, M07 is used to turn on the mist coolant. M09 turns off the coolant.

Automatic pallet changer

See Fig.4-4. An M60 command is commonly used to make pallet changes.

Fig.4-4 Machining center with APC

Other programmable features to look into

As stated, programmable functions will vary dramatically from one machine to the next. The actual programming commands needed will also vary from builder to builder[7]. Be sure to check the M code list (miscellaneous functions, refer to Appendix E) given in the machine tool builder's manual to find out more about what other functions may be programmable on your particular machine. M codes are commonly used by the machine tool builder to give the user programmable ON/OFF switches for machine functions. In any case, you must know what you have available for activating within your CNC programs.

For turning centers, for example, you may find that the tailstock and tailstock quill are programmable. The chuck jaw open and close may be programmable. If the machine has more than one spindle range, commonly the spindle range selection is programmable. And if the machine has a bar feeder, it will be programmable. You may even find that your machine's chip conveyor can be turned on and off through programmed commands. All of this, of course, is important information to the CNC programmer.

TECHNICAL WORDS

makeup	[ˈmeikʌp]	*n.*	组成，结构
utilize	[ˈju:tilaiz]	*v.*	利用
component	[kəmˈpəunənt]	*n.*	部件，元件
consistent	[kənˈsistənt]	*a.*	一致的
slant	[slɑ:nt]	*a.& v .*	倾斜的；倾斜
turret	[ˈtʌrit]	*n.*	转塔刀架
regarding	[riˈgɑ:diŋ]	*prep.*	关于
manual	[ˈmænjuəl]	*a. & n.*	手动的；手册，说明书
coordinate	[kəuˈɔ: dineit]	*n.*	坐标
sync	[siŋk]	*n.*	同步
probe	[prəub]	*n.*	探针，测头，对刀仪

document	['dɔkjumənt]	*v. & n.*	证明，记录；文件，资料
offset	['ɔːfset]	*n.*	偏置
miscellaneous	[misə'leiniəs]	*a.*	混杂的
activate	['æktiveit]	*v.*	激活，开启

PHRASES

conventional machine		普通机床
vary from...		随……而不同
in...manner		以……方式
turning center		车削中心
machining center		加工中心
rough turning		粗车
finish boring		精镗
necking		凹槽加工
motion control		运动控制
ball screw		滚珠丝杠
linear scale		直线光栅尺
table vise		台虎钳
work holding device		工件夹持装置，夹具
servo motor	(drive motor)	伺服电动机
assembly drawing		装配图
dovetail ways		燕尾导轨
coordinate system		坐标系
reference point		参考点
zero return		回参考点，回零点
grid zero		栅格零点
power up		上电，通电
machine tool builder	(MTB)	机床厂，机床制造商
post process gauging system		后处理测量系统
adaptive control system		自适应控制系统
live tooling		径向车铣动力刀座
constant surface speed	(CSS)	恒表面速度切削
automatic pallet changer	(APC)	自动托盘交换装置
automatic tool changer	(ATC)	自动刀具交换装置
spindle range		主轴挡位，主轴齿轮级

cut-off point	指主轴速度的切换点，换挡速度
miscellaneous function	M 功能
chip conveyor	排屑装置
bar feeder	送棒料装置

NOTES

1. The first goal of any CNC beginner should be to understand the basic machining practice that goes into using the CNC machine tool. The more the beginning CNC user knows about basic machining practice, the easier it will be to adapt to CNC. 数控初学者的第一个目标是理解数控机床使用过程中所需的基本的机械加工实践，初学者对基本的机械加工实践懂得越多，则越容易适应数控加工工作。

2. If you already know basic machining practice **as** it relates to the CNC machine you will be working with, you already know what **it** is you want the machine to do. 因为是否懂得机械加工与将来能否用好数控机床关系很大，所以，如果你已经懂得基本的机械加工实践，你就知道想要机床做什么。*这里 as 表示原因，主句中的 it 是形式主语，其真正的内容是 you want the machine to do。*

3. Since this form of CNC machine can perform multiple operations in a single program (as many CNC machines can), the beginner should also know the basics of how to process **workpieces machined by turning** so a sequence of machining operations can be developed for **workpieces to be machined**. 由于这种数控机床能在一个程序中完成多种加工（很多数控机床都能够如此），初学者同时应该了解如何加工车削类零件的基础知识，这样才能编制出待加工零件的加工工序。*workpieces （which are） machined by turning 是“通过车削加工的零件”的意思，workpieces （which are）to be machined 是“要被加工的零件”的意思。*

4. Rather than applying completely mechanical devices to cause motion **as** is required on most conventional machine tools, CNC machines allow motion control in a revolutionary manner. 大多数普通机床完全运用机械装置实现其所需的运动，而数控机床是以一种全新的方式控制机床的运动。*这里关系代词 as 引起定语从句，指代 applying completely mechanical devices to cause motion，在定语从句中作为主语。下文有类似的用法。*

5. A CNC command executed within the control tells the drive motor to rotate a precise number of times. The rotation of the drive motor in turn rotates the ball screw. And the ball screw drives the linear axis (slide). A feedback device (linear scale) on the slide allows the control to confirm that the commanded number of rotations has taken place. 数控系统中的 CNC 指令命令驱动电动机旋转某一精确的转数，驱动电动机的旋转随即使滚珠丝杠旋转，滚珠丝杠将旋转运动转换成直线轴（滑台）运动。滑台上的反馈装置（直线光栅尺）使数控系统确认指令转数已完成。

6. CNC controls that utilize a reference point for each axis **require** that the machine be manually sent to its reference point in each axis as part of the power up procedure. 各轴采用参考点的数控系统需要手动将机床的各轴返回参考点，这是上电过程中必不可少的部分。*这里 require 后跟的从句为虚拟语气。以 suggest, insist, demand, recommend, command, require, desire, propose 等动词为主句谓语时，从句须用虚拟语气，使用动词原形或 should+动词原形。*

7. The actual programming commands **needed** will also vary from builder to builder. 实际所需的编程指令也随制造厂家的不同而不同。*这里 needed 是过去分词作为后置定语。*

PRACTICE

Task 1 Translate the following phrases into English，using the gerund form of the verb.

1. 平面铣削	2. 轮廓铣削
3. 精镗加工	4. 槽铣削
5. 粗车加工	6. 切槽加工
7. 车螺纹	8. 浅孔钻

Task 2 Write the full name of the following abbreviations.

1. HP____________________________	2. RPM___________________________
3. MTB____________________________	4. CNC___________________________
5. ATC ____________________________	6. APC___________________________
7. CSS____________________________	8. HSS___________________________

Task 3 Write the correct form of each word as requested.

rotate______________________ (*n.*)	measure__________________ (*n.*)
automation ________________ (*a.*)	axis______________________ (*pl.*)
accurate____________________ (*n.*)	forward___________________ (*anto.*)
precise_____________________ (*n.*)	production________________ (*v.*)
specific_____________________ (*n.*)	vibrate___________________ (*n.*)
lubrication__________________ (*v.*)	mechanism________________ (*a.*)

Task 4 Fill in the brackets with words that have similar meaning to the underlined words, changing their forms if necessary.

1. (   ) These axes can be precisely and automatically positioned along their lengths of <u>travel</u>.

2. (   ) Rather than applying completely mechanical devices to cause motion as is required on most conventional machine tools, CNC machines allow motion control in a revolutionary <u>manner</u>.

3. (   ) The motion type (rapid, linear, and circular), the axes to move, the amount of <u>motion</u> and the motion rate (feed rate) are programmable with almost all CNC machine tools.

4. (   ) What is the machine's <u>rapid</u> rate?

5. (   ) What is the maximum <u>travel distance</u> in each axis?

6. (   ) The CNC programmer must know the programmable <u>motion directions</u> available for the CNC machine tool.

7. ( ) The actual programming commands needed will also vary from builder to builder.

8. ( ) Some control manufacturers call this position the zero return position.

9. ( ) Most CNC machines utilize a very accurate position along each axis as a starting point or reference point for the axis.

10. ( ) Information regarding the machine's construction including assembly drawings is usually published right in the machine tool builder's manual.

11. ( ) Information regarding the machine's construction including assembly drawings is usually published right in the machine tool builder's manual.

12. ( ) Information regarding the machine's construction including assembly drawings is usually published right in the machine tool builder's manual.

13. ( ) Information regarding the machine's construction including assembly drawings is usually published right in the machine tool builder's manual.

14. ( ) What is the machine's maximum RPM?

15. ( ) A feedback device on the slide allows the control to confirm that the commanded number of rotations has taken place.

Task 5 Decide whether the following statements are true (T) or false (F).

1. ( ) The CNC machine can perform multiple operations in a single program.

2. ( ) There is no need to learn about the basic machining practice, since the CNC machine is fully automated.

3. ( ) A machining center includes the bed, way system, headstock & spindle, turret construction, tailstock, and work holding device.

4. ( ) Once the machine is manually sent to its reference point in each axis as part of the power up procedure, the control will be in sync with the machine's position.

5. ( ) Each CNC machine has the same programmable functions.

6. ( ) Generally speaking, the more axes, the more complex the machine.

Task 6 Fill in the blanks with the appropriate form of the underlined words which can be of the different form, or synonym, or antonym, etc.

Example: CNC is a form of programmable automation in which the machine tool is controlled by a ________ in computer memory. The correct answer should be ***program.***

1. M03 is used to turn the spindle on in a clockwise manner. M04 turns the spindle on in a ____________________manner.

2. Common axis names are X, Y, and Z for linear axes and A, B, and C for __________ axes.

3. Common axis names are X, Y, Z, U, V, and W for linear __________and A, B, and C for rotary axes.

4. M06 tells the machine to actually make the tool change and ATC device is the automatic tool ________________on the machining center.

5. After cutting and finishing operations, a __________________part is formed.

6. A machinist is a person who uses machine tools to make or modify parts, primarily metal parts, a process known as ___________.

7. The axes can be precisely and automatically positioned along their lengths of travel. CNC controls require that the machine be ___________ sent to its reference point in each axis as part of the power up procedure.

8. Automobile parts, machine parts and compressors are precision products. They are cut and shaped by using CNC machines which are extremely ___________.

9. Tighten the chuck using the chuck key to ensure a tight and even grip. Rotate the workpiece to ensure that it is seated ___________.

10. If it is desired to increase either the feed or the depth of cut, the cutting speed should be proportionally ___________ to prevent overheating and excessive cutter bit wear.

Task 7 Fill in the blanks with words or phrases from the reading that match the meanings of the sentences, changing their forms if necessary. The first letters are already given. Then compare with your partner.

1. With the capabilities of maintaining c___________ at the point of the cutting tool, the spindle speed will automatically increase as the diameter decreases during a turning operation.

2. C___________ are features of machining centers that are used to help manage the huge amount of chips and debris these machines can produce in a short period of time.

3. C___________ is movement in two or more axes at the same time in order to produce a smooth continuous surface or curve.

4. Generically f___________ are used to hold the workpiece.

5. L___________ is powered tooling, such as a drill that may be held in the turret of a lathe.

6. Machine r___________ means movement along an axis in the farthest possible positive direction.

7. T___________ is the component of a lathe that holds a number of cutting tools. It's similar to a tool changer on a mill.

8. V___________ is a mechanical device used to clamp work for hand or machining operations.

9. R___________ means to move at the machine's maximum rate of feed in order to reposition the cutter to a new location prior to beginning a new cut. It can be many times faster than the cutting feed rate.

10. B___________ are highly efficient low friction and low backlash lead screw devices that use ball bearings rolling in a channel cut into the screw. They are used to drive the axes of the CNC machine.

Task 8 The following lists the technical data of a CNC lathe as shown in Fig.4-7. Find the answers to the questions raised in the text according to the following data.

Features

- It is equipped with SIEMENS 802C CNC with closed-loop control.

- Turning, boring, drilling, tapping and thread cutting can be performed.
- Slant bed construction with PTFE square guide ways（贴塑导轨）ensures good rigidity（刚性） and easy chip conveying.
- High quality bearing in the spindle head and high precision ball screws are employed.
- The machine has two spindle ranges.

Specifications

• Max. swing（回转直径）over bed	Φ500mm
• Max. swing over carriage	Φ282mm
• Max. turning diameter	Φ250mm
• Max. turning length	650mm
• Spindle speed	8-4000RPM
• Spindle range cut-off point	700RPM
• Max. spindle torque	533Nm
• Spindle motor horsepower	11/16kW
• Servo motor torque	X-axis: 6.8Nm; Z-axis: 6.8Nm
• Travels	X-axis: 250mm; Z-axis: 700mm
• Rapid traverse rates	X-axis: 20m/min; Z-axis: 20m/min
• Fastest cutting feed rate	2m/min
• Tool turret	8 positions

Task 9 Scan reading the following passage, and then select an answer from the four (or three) choices.

Each machine relies on Electronics Industries Association (EIA) standards that dictate the location of the axes. The following are general guidelines that determine the location of coordinate axes on any given CNC machine:

- The Z-axis is always parallel to the spindle of the machine.
- The X- and Y-axes are always perpendicular to the spindle of the machine.
- The X-axis normally describes the longer direction of travel on the machine, and the Y-axis describes the shorter direction of travel.

Coordinates for the vertical milling machine

The most recognizable coordinates can be found on the vertical milling machine. Imagine that you are standing in front of the machine and facing it. As you can see in Fig.4-5, the X-axis describes left and right motion of the cutting tool, the Y-axis describes back and forth motion of the tool, and the Z-axis describes up and down motion.

Depending on the machine, either the cutting tool or the worktable will move during machining operations. The positive and negative directions on each axis always describe the motion of the cutting tool in relation to the worktable.

Coordinates for the horizontal milling machine

Fig.4-6 shows the basic setup of a horizontal milling machine. As you can see, the spindle is located on the side. The Z-axis must always be parallel to the spindle of the machine. The X-axis still describes motion to the left and right. However, the Y-axis now describes motion that is up and down.

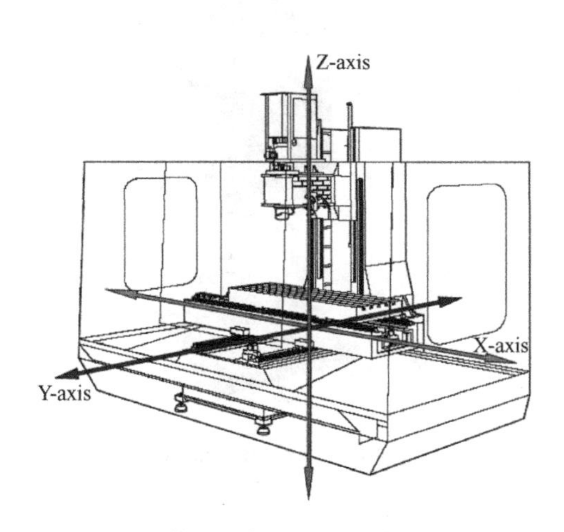

Fig.4-5 Coordinates for the vertical milling machine

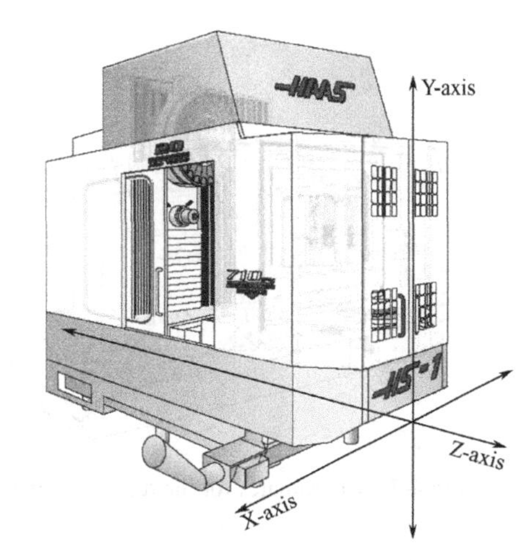

Fig.4-6 Coordinates for the horizontal milling machine

Coordinates for the turning center

Milling machines are normally used to machine cubic（立方体的） workpieces. However, manufacturers will use turning centers to shape the dimensions of cylindrical workpieces.

During turning operations, the workpiece is held and rotated in a spindle. A non-rotating cutting tool is moved against the rotating part to remove material. Because the Z-axis is always parallel to the spindle of the machine, it no longer describes the location of the cutting tool. Instead, the Z-axis now describes the back and forth motion of the tool along the length of the workpiece. The X-axis specifies the distance of the cutting tool from the center of the workpiece. Fig.4-7 shows both the X- and Z-axes.

The X-axis position of the cutting tool determines the diameter of the cylindrical workpiece. On the typical turning center, the Y-axis is not programmable. In theory, the Y-axis defines the tool height. However, because the tip of the tool must cut on the centerline of the part, the tool is almost always fixed at the same Y-axis location.

Contouring

Not all cutting operations take place in either a vertical or horizontal direction. A CNC machine will often be required to cut along a diagonal（对角的，斜的） line. This movement is called linear interpolation（插补）. To make this cut, a CNC machine moves on both the X- and Y-axes simultaneously（同时地）.

Now imagine that the same diagonal cut needed to be made, except that the cutting tool must gradually lift off the table as well. This cutting operation would require movement on the X-, Y-, and Z-axes simultaneously.

Sophisticated CNC machines can move a tool along two or more axes, either linear or rotational, at once. This movement is called contouring（轮廓加工）. See Fig.4-8. However, very few CNC systems factor all six axes at once during their operations, and many only address two or three axes at the same time.

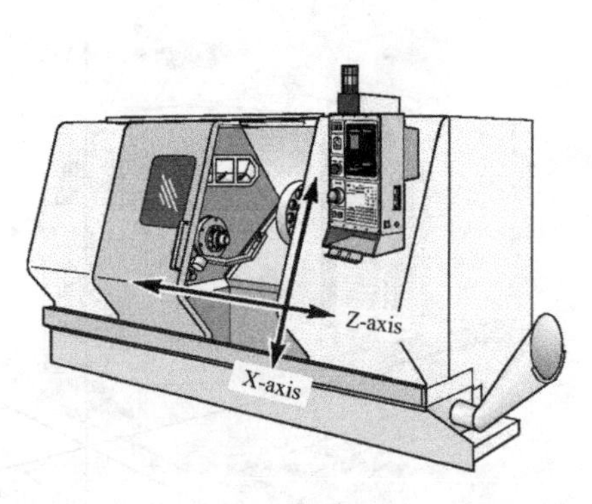

Fig.4-7 Coordinates for the turning center

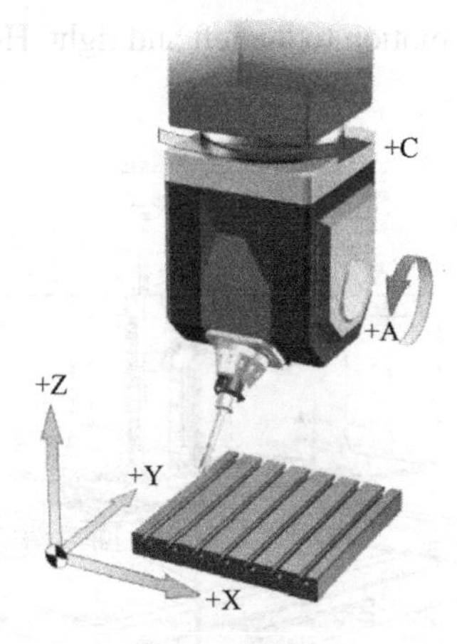

Fig.4-8 5-axis contouring

Questions:

1. On the vertical milling machine, the horizontal movement of the worktable will be in the:

(A) Y- and X-axes

(B) Y-axis

(C) Z- and Y-axes

(D) Z-axis

Your answer: ________

2. How are the X-, Y-, and Z-axes related?

(A) They are the same.

(B) They are parallel.

(C) They intersect to form a corner.

Your answer: ________

3. On most machines, the Z-axis corresponds to the:

(A) Worktable

(B) Work holder

(C) Spindle

(D) Part

Your answer: ________

4. On a vertical machining center, if the Z-axis is vertical, up is usually:

(A) Negative

(B) Positive

(C) Left

(D) Right

Your answer: ________

5. Coordinates used on the CNC machines:

(A) Move parts to new locations

(B) Move tools to new locations

(C) Specify precise locations

(D) Specify axis movement

Your answer: __________

6. On the horizontal milling machine, the horizontal movement of the worktable will be in the:

(A) Z- and X-axes

(B) X-axis

(C) Z- and Y-axes

(D) Z-axis

Your answer: __________

7. The Cartesian（笛卡儿）coordinate system describes:

(A) Locations using numbers

(B) Lines using numbers

(C) Lines

(D) Numbers

Your answer: __________

8. What linear axis is not programmable on a basic turning center?

(A) A-axis

(B) Z-axis

(C) Y-axis

(D) X-axis

Your answer: __________

9. Rotational axes:

(A) Are perpendicular to the linear axes

(B) Are parallel to the linear axes

(C) Go through the linear axes

(D) Go around the linear axes

Your answer: __________

10. What is the minimum number of axes that contouring moves through?

(A) Four

(B) Three

(C) Two

(D) One

Your answer: __________

Task 10 Find the missing words for the following passage and then read it aloud.

The two most common CNC machines are the ____________ and the ______________. A turning center is used to machine ____________ parts, and a machining center is used to machine flat or _____________ surfaces. For a universal style slant bed turning center, the programmer should know the most basic machine ______________, including the bed, way system, headstock & spindle, ______ construction, tailstock, work holding device and CNC control. For the machining center, it includes the bed, saddle, column, worktable, servo motors, ball screws, spindle, tool changer, and _____________. Information regarding the machine's construction including assembly drawings is usually published right in the ________________'s manual.

Part B Listening

Task 1 Listen to the five statements twice and write them down.

1. --
2. --
3. --
4. --
5. --

Task 2 The following audio introduces the technical data of a vertical machining center. Listen carefully and complete the list of specifications, paying attention to the measuring units.

Travels		
X-axis		
Y-axis		
Z-axis		
Feed rates		
Rapid traverse rates	X-axis	
	Y-axis	
	Z-axis	
Fastest cutting feed rates		
Spindle		
Spindle speed		
Max. torque		
Spindle power		
CNC control		
FANUC		
ATC		
Tool shank（刀柄）		
Tool magazine（刀库）capacity		
Tool change time		
Chip-to-chip time		
Accuracy		
Positioning（定位精度）		
Repeatability（重复定位精度）		

Task 3 Here are eight questions. Following each question, there is a video clip. Watch the video clips, then choose the best answer to the questions followed.

1. Which word or phrase best describes the operation that you see?

(A) Drilling.

(B) Boring.

(C) Reaming.

(D) Probing.

Your answer: ___________

2. Which word or phrase best describes the operation that you see?

(A) Face milling.

(B) Pocket milling.

(C) Drilling.

(D) Probing.

Your answer: ___________

3. Which word or phrase best describes the operation that you see?

(A) Face milling.

(B) Pocket milling.

(C) Surface grinding.

(D) Probing.

Your answer: ___________

4. Which word or phrase best describes the operation that you see?

(A) Face milling.

(B) Tapping.

(C) Surface grinding.

(D) Probing.

Your answer: ___________

5. What device can you see?

(A) An automatic tool changer.

(B) An automatic loader.

(C) An automatic pallet changer.

(D) A robot.

Your answer: ___________

6. What machine is shown in the video clip?

(A) An FMC.

(B) A vertical machining center.

(C) A horizontal machining center.

(D) A turning center.

Your answer: ___________

7. What you see is:

(A) Tool changing on a horizontal machining center.

(B) Tool changing on a vertical machining center.

(C) Pallet changing on a horizontal machining center.

(D) Pallet changing on a vertical machining center.

Your answer: ____________

8. Which word or phrase best describes the operation that you see?

(A) Thread milling.

(B) Pocket milling.

(C) Face milling.

(D) Contour milling.

Your answer: ____________

Task 4 The following video introduces the CNC turning center manufactured by Hardinge.Inc., a famous American MTB. Watch it first, then listen to it twice and fill in the following blanks with what you hear and see. (Refer to Part E)

(1)__________________is available on VDI （德国工程师协会）top plate to work on the main spindle or sub-spindle option（选择功能）. Each station can be equipped with a driven tool（动力刀具）for cross-or end-milling/drilling operations on the toughest materials. (2)______________ spindle orient is included. Internal and external coolant-style live tool holders are offered to direct (3)___________ to the work area. Angular drilling or milling is easily accomplished using adjustable VDI live tooling attachments at any station.

The (4)_________ option allows thread milling and complex off-center milling and drilling operations on the main or sub-spindle when using the live tooling option. The machine features an impressive (5)___________ overall Y-axis stroke（行程）.

The A2-5 sub-spindle option is offered in a belted or high speed wraparound（包缠式的）configuration with a thru-capacity（通过量）up to 15⁄8 "with 16C collets（夹头）and a gripping capacity of 51⁄2"with (6)________________short jaw chucks. Spindles have exact synchronization（同步）between the main and sub-spindles at any RPM for part transfer for secondary machining.

C-axis contouring provides positioning in increments（增量）of (7)____________ degree and is available on both the main and sub-spindle. With the (8)______________ option, you can perform three dimensional contouring, complex round and prismatic（棱镜的）machining, and lettering（印字）by synchronizing the spindle with the (9)____________ and (10)______________-axis.

Task 5 For certain applications, you need to choose your machine configuration with different options and standard features. Watch and listen to the video again and match Column A, B, and C.

No.	Standard features (Column A)	Options (Column B)	Applications (Column C)
1	12-station VDI 30 turret top plate	Live tooling option	Cross- or end-milling/drilling
2	10 or 12-station T-style top plate for static tooling	Y-axis option	Thread milling and complex off-center milling and drilling
3	Rigid tapping	Sub-spindle option	Part transfer for secondary machining
4	One-degree spindle orient	C-axis option	Three dimensional contouring
5	Graphic tool path display	Polygon turning option	Complex round and prismatic machining
6	Tool life management	PC front-end control option	Lettering
7	Variable lead thread cutting		Producing hexes, squares, triangles or flats in multiples of two or three
8	Run time and parts counter		

Task 6 Watch and listen to the above video once more, and choose the best answer to each of the following questions.

1. What is the brand name of the machine?

(A) HAAS.

(B) MAZAK.

(C) CINCINATI.

(D) HARDINGE.

Your answer: ___________

2. If you want to make cross- or end-milling/drilling, which options are essential according to the video?

(A) Live tooling option.

(B) Y-axis option.

(C) Sub-spindle option.

(D) C-axis option.

Your answer: ___________

3. If you want to make thread milling and complex off-center milling/drilling, which options are essential according to the video?

(A) Live tooling option and Y-axis option.

(B) Sub-spindle option.

(C) C-axis option.

(D) Polygon option.

Your answer: ___________

4. If you want to perform three dimensional contouring, complex round and prismatic machining, and lettering, which options are essential according to the video?

(A) Live tooling option.

(B) Y-axis option.

(C) Sub-spindle option.

(D) C-axis option.

Your answer: ___________

5. If you want to produce polygon shapes on the outside diameter of a part, which options are essential according to the video?

(A) Live tooling option and Y-axis option.

(B) Sub-spindle option.

(C) C-axis option.

(D) Polygon option.

Your answer: ___________

6. Which is a true statement about the CNC control according to the video?

(A) HARDINGE charges extra for features of graphic tool path display, rigid tapping and tool life management.

(B) Two USB ports are provided as the standard feature.

(C) Ethernet is provided as an option.

(D) The feature of variable lead thread cutting is optional.

Your answer: ___________

Part C Speaking

Task 1 Watch the slides and give the English name or description for each of the slides. Take notes.

1. ______________________ 6. ______________________

2. ______________________ 7. ______________________

3. ______________________ 8. ______________________

4. ______________________ 9. ______________________

5. ______________________ 10. ______________________

Task 2 Explain the following codes in English, using such verbs as specify, command, designate, or the structure like be used to.

1. M03 2. M05

3. M06 4. M07

5. F200 6. S900

7. G00 8. G01

9. G02 10. T0505

Task 3 Work in pairs. Use the following words to make a situational dialogue containing at least 5 sentences. Take notes.

CNC, machine tool, coolant, spindle, program, motion, ATC

Task 4 Watch the video about SIEMENS virtual machines and list words and phrases about them. After that, tell the class what's about the video just as a narrator.

1. ____________ 6. ____________
2. ____________ 7. ____________
3. ____________ 8. ____________
4. ____________ 9. ____________
5. ____________ 10. ____________

Task 5 Shop floor practice. Get to know the CNC machine which is new to you. Recognize the basic components and ask the teacher questions about it. Take notes.

Task 6 Work in groups. Create a situation of on-site demonstration and role-play the situation before the class. One acts as the machine operator as well as the narrator, some others as the visitors.

WORKING SITUATION

Some foreign guests visit your company. When the visitors come to your machine, you will be the best person to introduce the features of the machine. You need give an **on-site demonstration**.

Introduction to the machine

Part D Grammar and Translation

as 的用法

as 有多种用法，归纳如下。

1. 意思是"如，像，按"

as 是连词，引导方式状语从句，从句常常是省略句，如：

This form of CNC machine can perform multiple operations in a single program **as** many CNC machines can (=as many CNC machines can perform multiple operations in a single program). 这种数控机床能够在一个程序里完成多种操作，这就像很多其他的数控机床一样。

As stated (as it is stated), programmable functions will vary dramatically from one machine to the next. 如前所述，可编程功能因机床不同变化很大。

A selective feed lever, here called the longitudinal& cross feed lever, is provided for engaging the longitudinal feed or cross feed **as desired**. (Ref. LS 2) 提供进给选择手柄（这里称为纵向/横向进给手柄），用于按需进行纵向进给或者横向进给的齿轮啮合。

You must consider the rotational speed of the workpiece and the movement of the tool relative to the workpiece **as discussed** earlier. (Ref. LS 3) 如前面讨论的那样，必须考虑工件的旋转速度和刀具相对工件的移动速度。

2. 意思是"当……时候"

as 是连词，引导时间状语从句，如：

As you begin to learn about any new CNC machine, you should concentrate on four basic areas. 你在开始学习任何一台新的数控机床的时候，应该重点学习四个基本方面的内容。

Never attempt to break away metal spirals **as** they form at the cutting tool. (Ref. LS 3) 当金属螺旋状切屑在刀尖形成的时候，千万别去弄断它。

3. 意思是"由于，随着"

as 是连词，引导原因状语从句，如：

Job opportunities for machinists should continue to be good, **as** employers value the wide-ranging skills of these workers. (Ref. LS 1) 由于雇主们看重工人的全面技能，机械工的就业机会被持续看好。

If you already know basic machining practice **as** it relates to the CNC machine you will be working with, you already know what it is you want the machine to do. 因为是否懂得机械加工与你将来能否用好数控机床关系很大，所以，如果你已经懂得基本的机械加工实践，你就已经知道你想要机床做什么。

4. 意思是"……像……一样"

as...as...，第一个 as 是副词，第二个 as 是连词，引导比较状语从句，如：

Keep tools overhang **as** short **as** possible. (Ref. LS 3) 保持刀具突出部分尽可能短。

You want the workpiece to be **as** parallel **as** possible with the center line of the lathe spindle. (Ref. LS 3) 工件应尽可能与车床主轴中心线平行。

5. 意思是“作为”

as 是介词，如：短语 be known as，refer to...as。又如：

Most CNC machines utilize a very accurate position along each axis **as** a starting point or reference point for the axis. 大多数数控机床的各个坐标轴都采用一个很精确的位置作为各轴的起始点或参考点。

An interlocking device, that prevents the half-nuts and the feed mechanism from engaging at the same time, is usually provided **as** a safety feature. (Ref. LS 2) 通常提供互锁装置作为安全功能，防止开合螺母和进给机构同时啮合。

6. 意思是“所……的”

as 是关系代词，引导定语从句，如：

Instead of causing motion by turning cranks and handwheels as is required on conventional machine tools like engine lathes, CNC machines allow motions to be commanded through programmed commands. 像车床这样的普通机床须通过旋转摇柄和手轮产生运动，而数控机床则通过编程指令产生运动。

The lathe or engine lathe, as the horizontal metal-turning machine is commonly called, is considered the father of all other machine tools. (Ref. LS 2) 卧式金属车床常称为车床或普通车床，被认为是机床之父。

Task Translate the following sentences into Chinese.

1. The headstock is required to be made as robust as possible due to the cutting forces involved.

----

2. There is also the potential that the work could be forced to loosen in the chuck jaws and fly out as a dangerous projectile.

----

3. As always, wear safety glasses and keep your face well away from the work since this operation will throw off hot chips and/or sharp spirals of metal.

----

4. As the tool starts to cut into the metal, maintain a steady cranking motion to get a nice even cut.

----

5. The cutting speed of a tool bit is defined as the distance of the workpiece surface, measured at the circumference, that passes the tool bit in one minute.

Part E Supplementary Reading

HARDINGE's QUEST® lathe

Ideal for basic lathe machining for multi-tasking operations, HARDINGE's QUEST® lathes (Fig.4-9) are offered in both general precision and super precision models. Choose your exact machine configuration with a wide range of multi-tasking options to quickly and efficiently produce precise and complex parts in a single setup. Here we will demonstrate the multi-tasking features offered in HARDINGE's QUEST® lathes.

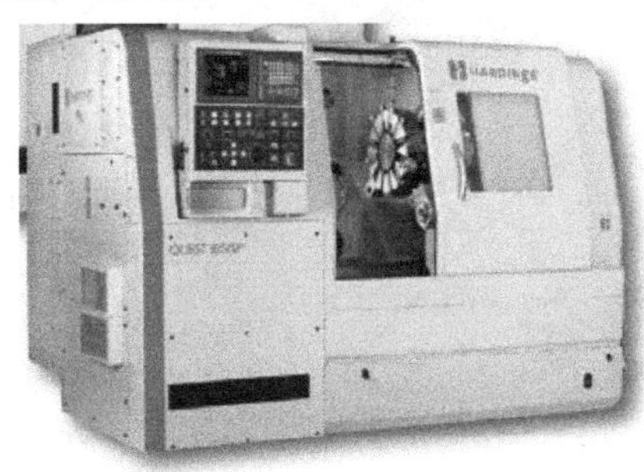

Fig.4-9 HARDINGE's QUEST® lathe

Features are designed to help you maximize your productivity. Choose from a standard 12-station VDI 30 turret top plate with or without live tooling（动力刀座） and Y-axis options（选项功能）, or a HARDINGE 10 or 12-station T-style top plate for static tooling compatibility with HARDINGE's T-series lathes. Fig.4-10 shows the 12-station VDI 30 turret top plate.

Rigid tapping（刚性攻丝）capability is standard on all QUEST® lathes. With rigid tapping, you will eliminate the needs for tension and compression tap holders.

Live tooling is available on VDI top plate（VDI 德式刀架）to work on the main spindle or sub-spindle option as shown in Fig.4-11. Each station can be equipped with a driven tool for cross- or end-milling/drilling operations on the toughest materials. One-degree spindle orient（主轴定向） is included. Internal and external coolant-style live tool holders are offered to direct coolant to the work area. Angular drilling or milling is easily accomplished using adjustable VDI live tooling attachments（附件） at any station.

Fig.4-10 Turret top plate with live tooling and the axes

Fig.4-11 Main, sub-spindle and the C-axis

The Y-axis option allows thread milling and complex off-center milling and drilling operations on the main or sub-spindle when using the live tooling option. The machine features an impressive 3.38" overall Y-axis stroke（行程）. See Fig.4-12.

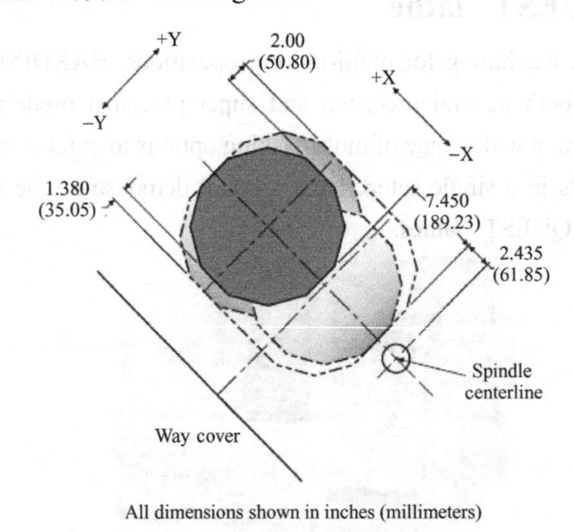

Fig.4-12 Y-axis stroke

The A2-5 sub-spindle option is offered in a belted or high speed wraparound（包缠式的）configuration with a thru-capacity（通过量）up to 1⅝" with 16C collets（夹头）and a gripping capacity of 5½" with 6" short jaw chucks. Spindles have exact synchronization（同步）between the main and sub-spindles at any RPM for part transfer for secondary machining.

C-axis contouring provides positioning in increments of .001 degree and is available on both the main and sub-spindle. With the C-axis option, you can perform three dimensional contouring, complex round and prismatic（棱镜的）machining, and lettering（印字）by synchronizing the spindle with the X and Z-axis. See Fig.4-11.

With the polygon（多边形）turning option, you can produce polygon shapes on the outside diameter of a part in a single Z-axis move. This option is recommended for producing hexes（六角形）, squares（四方形）, triangles（三角形） or flats in multiples of two or three. Polygon turning can be performed on either the main or sub-spindle. As shown in Fig.4-13, this highly productive method of producing wrench flats（扳手平面部分）on a part is accomplished by synchronizing the parts spindle with the live tooling spindle. The cutter shown here has three cutting inserts, which allows multiples of three flats. As the cutter and the part rotate, each tool cuts a flat. To accomplish a part with six flats, you'll simply run the cutter twice the speed as the parts spindle. An impressive feature of this type of machining is the ability to chamfer the flats, or make flats on a taper.

Fig.4-13 Polygon turning

HARDINGE's QUEST® lathes feature a custom-designed（用户定制设计的）CNC control with many standard features other machine tool builders charge extra for. See Fig.4-14. Features such as graphic tool path display, rigid tapping, tool life management, variable lead thread cutting, and run time and parts counter. The PC front-end control option includes a CD drive, 3.5" floppy drive（软盘驱动器）, Ethernet（以太网）, two USB ports, and Microsoft Windows NT operating system.

Fig.4-14 Custom-designed CNC control

A host of multi-tasking features and options make the HARDINGE's QUEST® lathes the right machines for your shop today and for years to come.

Learning Situation 5

Learn the HAAS control panel

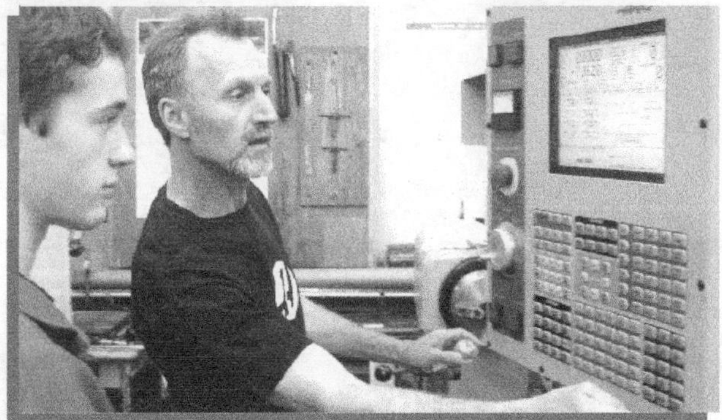

Focus of the situation

This class will introduce you to HAAS mill control panel components, teach you machine control keypad functions, and familiarize you with powering up and powering down the HAAS mill. You will also learn some basic operations, such as homing the machine, activating the chip conveyor, checking coolant levels, activating coolant, and leaving messages. [本课介绍HAAS 铣床控制面板的组成、键盘的功能，教你进行机床的通电和断电、参考点的返回、冷却液的开关、排屑装置的开关等操作。]

Field work

Perform the above mentioned operations at a HAAS machining center with the help of the text.

Part A Reading

Introduction to HAAS control

The HAAS control is one of the most common types of CNC controls in the shop today. While other CNC mills may have machines and controls from different manufacturers, every HAAS mill comes with a HAAS control[1]. Once you learn its specifics, you will find that the HAAS control is logical and easy to use.

The control panel

You control all machine operations from the control panel. The control panel, as shown in Fig.5-1, consists of manual controls, the display screen, and the control keypad:

- POWER ON and POWER OFF buttons turn on and shut down the machine.
- The SPINDLE LOAD meter indicates the power draw on the spindle motor. It tells you how hard your machine is working and helps you keep your machine from being overworked and damaged.
- The EMERGENCY STOP button automatically shuts down all machine functions. You should use this button if a tool is about to collide with a part or fixture.
- The HANDLE moves the machine components along the axes. The handle "clicks" in controlled, measured increments that you select on the keypad.
- The CYCLE START button begins or restarts a program that had been stopped, and the FEED HOLD button stops axis motion but continues spindle movement. To turn off the spindle, press the SPINDLE STOP override key. FEED HOLD allows you to check the part or tooling.
- The display screen shows you the program, axis locations, and other relevant information throughout the machining process[2].
- The control keypad allows you to enter commands, enter offsets, and adjust overrides.

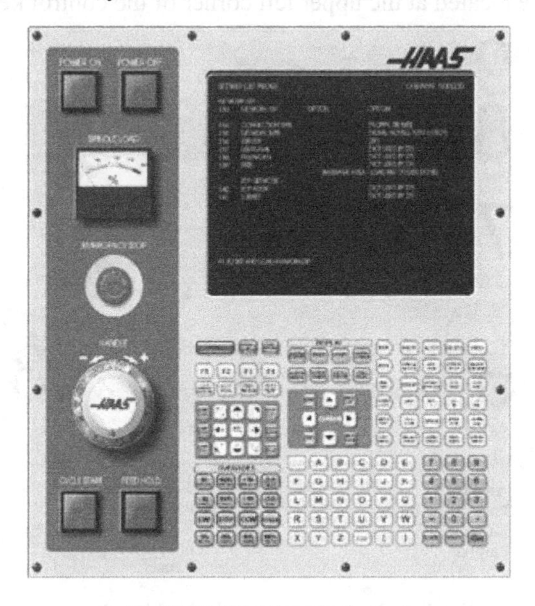

Fig.5-1 HAAS control panel

Fig.5-2 and Fig.5-3 show manual controls of the HAAS mill. Manual controls such as HANDLE, EMERGENCY STOP, CYCLE START, and FEED HOLD function much like the controls on other machines[3].

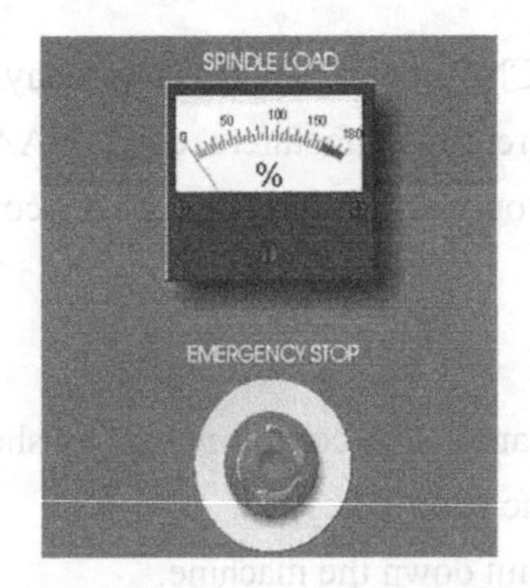

Fig.5-2 Spindle load meter and emergency stop button

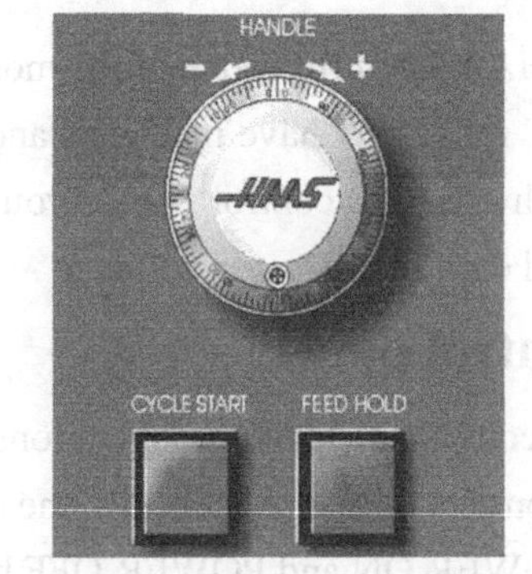

Fig.5-3 Handle, cycle start and feed hold button

The control keypad

The control keypad is where you enter commands for the HAAS control. As shown in Fig.5-4, there are nine areas of the control keypad:

- Eight display keys are located at the top center portion of the control keypad.
- Thirty operation mode keys are located at the upper right corner of the control keypad.
- Fifteen numeric keys are located at the lower right corner of the control keypad.
- Thirty alpha keys are located at the bottom center portion of the control keypad.
- Fifteen overrides are located at the bottom left corner of the control keypad.
- Eight cursor arrow keys are located at the center of the control keypad.
- Fifteen jog keys are located in the section above the overrides.
- Eight function keys are located above the jog keys.
- Three reset keys are located at the upper left corner of the control keypad.

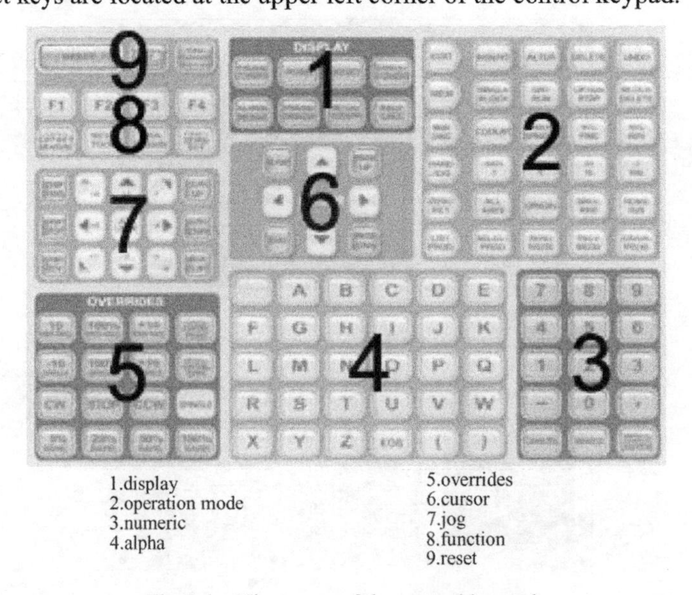

Fig.5-4 Nine areas of the control keypad

Display keys - top row

The display keys are located at the top center portion of the keypad. Pressing each key shows different information, such as offsets, machine settings, and the current program that is running. Some of these keys show another screen and additional information when pressed twice. The upper left corner of the display screen shows the current display. The top row of the display keys, as shown in Fig.5-5, provides essential information needed to successfully run a program:

- Pressing PRGRM/CONVRS shows the program blocks of the current selected part program. Pressing PRGRM/CONVRS twice allows an operator to enter codes for conversational programming in edit mode.
- Pressing POSIT shows five different pages that list machine axis positions of the spindle.
- Pressing OFFSET shows the various offsets.
- Pressing CURNT COMDS shows the fifteen lines of the current program, modal program values, and the position during run time.

Depending on the task, you will want to choose a specific display. When you load new tools and enter offsets, you can choose the OFFSET display. When you are running a job, choose the PRGRM/CONVRS or CURNT COMDS displays.

Display keys - bottom row

The bottom row of the display keys, as shown in Fig.5-6, provides you with additional displays and information. Like the top row of keys, some of the display keys in the bottom row show another screen when pressed twice:

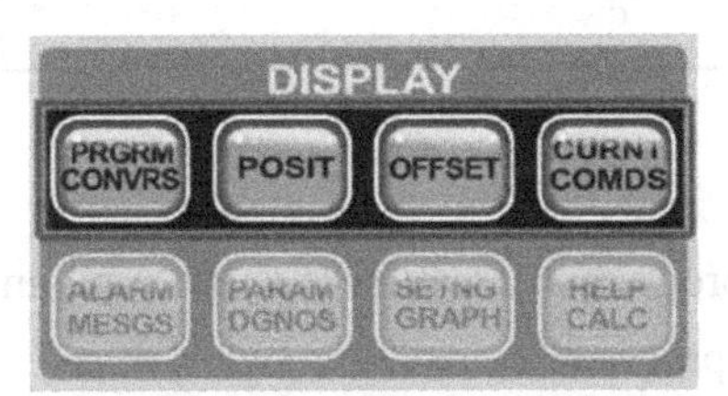

Fig.5-5 Top row of the display keys

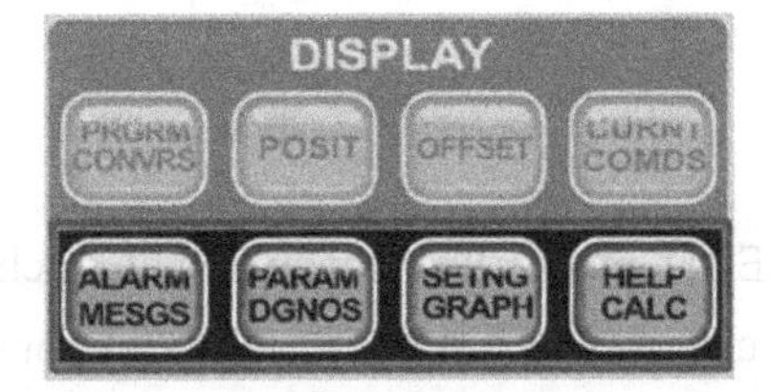

Fig.5-6 Bottom row of the display keys

- Pressing ALARM/MESGS once shows active alarms. Pressing the right cursor arrow in the alarm display shows the alarm history, while pressing the right cursor arrow twice shows alarm descriptions. Pressing ALARM/MESGS twice shows a message either for you or for the next operator.
- Pressing PARAM/DGNOS once lists machine parameters, which rarely change. HAAS advises that you not change parameters unless you know exactly what needs to be changed and why, and that you have made all the correct inquiries within your shop and with HAAS service personnel. If a parameter is changed without proper assistance, you may void the warranty of the machine[4]. Pressing PARAM/DGNOS twice shows the diagnostics display pages.
- Pressing SETNG/GRAPH once shows settings, which are machine control functions you may need to activate, deactivate, or change to suit specific conditions[5]. You will learn particular settings later in this class. Pressing SETNG/GRAPH twice allows you to run a

part program visually without risking tool or machine damage due to programming errors[6].

- Pressing HELP/CALC once shows an information manual, which contains 26 Help topics. Pressing HELP/CALC twice lets you choose between a TRIG calculator, CIRCULAR calculator, MILLING/TAPPING calculator, CIRCLE-LINE TANGENT calculator, or a CIRCLE-CIRCLE TANGENT calculator.

Operation modes

As shown in Fig.5-1, the operation mode keys are in the upper right corner of the control keypad. An operation mode tells what tasks the control will perform. The control can be in only one mode at a time. You can choose a particular mode from the buttons in the left column, as shown in Fig.5-7.

The keys in the same row as the pressed mode key are available after a mode has been selected[7]. As shown in Fig.5-8, the current mode is displayed on the top line of the display screen, just to the right of the current display. There are six operation modes:

Fig.5-7 Operation mode keys

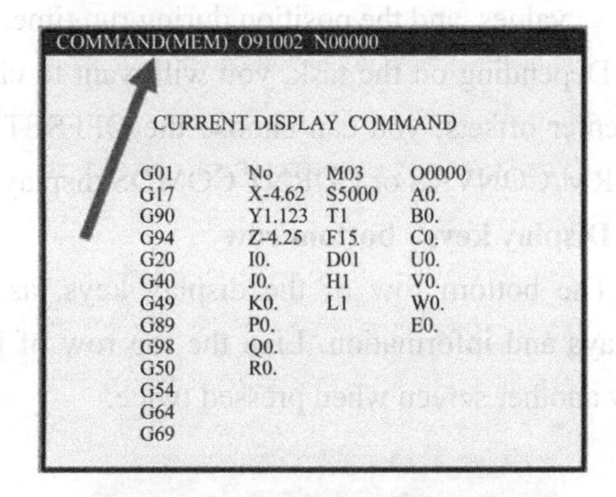

Fig.5-8 Current mode display

- EDIT mode allows manual editing changes in a program or creation of a new program. You may INSERT, ALTER, DELETE, or UNDO any program.
- MEM mode runs part programs from the control's memory. Pressing MEM shows the current program which will start when you hit CYCLE START.
- MDI/DNC mode places the machine in Manual Data Input mode. MDI lets you enter and execute program data without disturbing the stored programs. Pressing MDI/DNC twice activates Direct Numerical Control (DNC) if Setting 55 is on. DNC allows you to execute programs sent from a floppy drive or a computer hard drive.
- HANDLE JOG mode allows you to move the spindle along its axes with either the HANDLE or JOG buttons[8].
- ZERO RET mode searches for machine zero or rapid returns to machine zero automatically.
- LIST PROG mode lists programs and allows you to select, send, receive, and delete programs. To create a new program, you must be in the PRGRM/CONVRS display and LIST PROG mode. Enter **Onnnnn** from the keyboard and press SELECT PROG button. Press EDIT to show the new program.

Alpha keys and numeric keys

The numeric keys, as shown in Fig.5-9, are in the lower right corner of the keypad and allow you to enter numbers (0-9), a minus sign (-), and a decimal (.).

The alpha keys, as shown in Fig.5-10, provide you with all the capital letters of the alphabet (A-Z). Lower case letters (a-z) can be entered in between parentheses if you press or hold the SHIFT key and then press the letter.

Fig.5-9 Numeric keys

Fig.5-10 Alpha keys

Special characters

Fig.5-11 and Fig.5-12 show a variety of special characters. In particular, each of these special characters has a certain meaning in the program:

&	@	:
%	$	!
*	,	?
+	=	#

Fig.5-11 Special characters in the upper left corner

End of Block ;
Block Delete /
Text Information ( )
Macro Statements [ ]

Fig.5-12 Special characters of the alpha keys

Overrides

The overrides, shown in Fig.5-13, are located at the lower left of the control keypad. They give you the ability to alter the programmed feed and speed, spindle direction, and rapid traverse motion. By using overrides, you can adjust these variables "on the fly" while the program is being executed[9].

-10 FEED RATE decreases current feed rate in increments of 10%, and +10 FEED RATE increases the feed rate 10% with each button push. 100% FEED RATE sets the feed rate back to the programmed value. HANDLE CONTROL FEED allows you to turn the HANDLE to adjust the programmed feed rate at 1% increments if Setting 104, Jog Handle to Single Block, is OFF.

-10 SPINDLE decreases current spindle speed in increments of 10%, and +10 SPINDLE increases speed by 10% with each button push. 100% SPINDLE sets the spindle speed at the programmed value. HANDLE CONTROL SPINDLE allows the HANDLE to control spindle speed at 1% increments if Setting 104 is OFF.

In the last two rows, CW starts the spindle in the clockwise direction. STOP stops the spindle. CCW starts the spindle in the counterclockwise direction. 5% RAPID, 25% RAPID, and 50% RAPID allow you to reduce the speed of rapid travel by a specific percentage. 100% RAPID tells the machine to rapid traverse at its maximum rate.

Cursor keys

Shown up close in Fig.5-14, the cursor arrow keys are also used to edit and search for CNC programs. By using the arrow keys, you can move the cursor up, down, left, and right through the program or screen options.

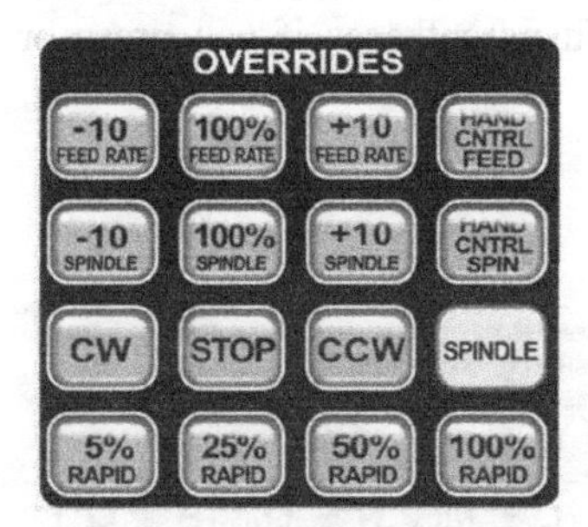

Fig.5-13 Overrides

Fig.5-14 Cursor keys

Jog keys

The jog keys, which are shown in Fig.5-15, are located in the section on the left next to the cursor arrow keys. If you hold down a jog key, it provides continuous motion of the spindle along the selected axis. The +X and -X buttons move the spindle along the X-axis. The +Y and -Y buttons move the spindle along the Y-axis. The +Z and -Z buttons move the spindle along the Z-axis. Pressing SHIFT and either +A or -A gives you access to the B-axis. When you press the JOG LOCK key before one of the above keys, the axis is moved in a continuous motion without the need to hold the axis key depressed[10]. Another press of the JOG LOCK key stops jogging motion.

In the left column next to the jog keys are buttons that control the chip auger. CHIP FWD turns the chip auger to remove chips from the machine. CHIP STOP stops chip auger movement, and CHIP REV turns the chip auger in the reverse direction.

In the column to the right of the jog keys are buttons that control the coolant for the machine. If coolant spigots are enabled, CLNT UP positions the coolant stream direction one position higher, the CLNT DOWN positions the coolant stream direction one position lower, and AUX CLNT turns on the Through the Spindle Coolant system if in MDI mode. Pressing AUX CLNT twice shuts off the system.

Function keys

The eight function keys, which are shown in Fig.5-16, are located immediately above the jog keys. Function keys perform a variety of different tasks. Some always perform the same task. Other function keys depend upon which display is currently viewed in the display screen.

Fig.5-15 Jog keys

Fig.5-16 Function keys

In the top row, F1, F2, F3, and F4 perform different functions depending on what display and mode are selected. These keys are located immediately below the RESET key and are used with editing, graphics, background edit, and the help/calculator to execute special functions.

The bottom row contains four keys that execute specific tasks:

- TOOL OFSET MESUR sets the present Z-location for the tool length offsets in the offset page during part setup[11].
- NEXT TOOL is used to activate the tool changer and select the next tool during part setup.
- TOOL RELEASE releases the tool from the spindle when in MDI, handle jog, or zero return mode. The tool release button must be held for one-half second before the tool is released and the tool will remain released for one-half second after the button is released.
- PART ZERO SET is used to enter work coordinate offsets automatically during part setup.

Reset keys

The reset keys are located in the upper left corner of the control keypad. Fig.5-17 shows the location and closeup view of these reset keys. The red RESET button stops all machine motion and places the program pointer at the top of the current program.

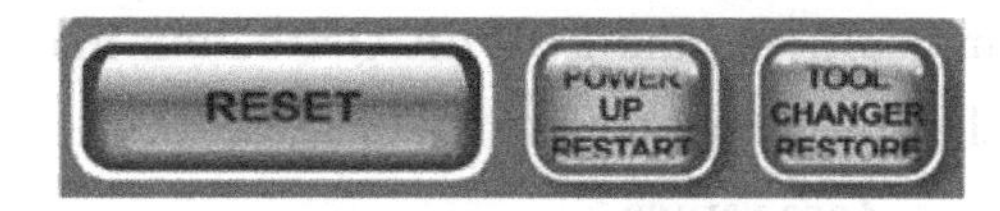

Fig.5-17 Reset keys

The POWER UP/RESTART button automatically initializes the machine at power up. After initial power up, this key can be used to reinitialize the system.

The TOOL CHANGER RESTORE button restores the tool changer to normal operation if the tool changer has encountered an interruption during a tool change. The key initiates a user prompt screen to assist the operator in recovering from a tool changer crash[12].

Basic operations

Now that you are familiarized with the CNC control and the control keypad, you may power up the machine. You may even perform some basic operations that lead up to running a part creation cycle.

Powering up

Powering up a HAAS mill is accomplished in a few easy steps, which are illustrated in Fig.5-18.

1. Press the green button labeled POWER ON. This button is located in the upper left corner of the control panel and is shown up close in Fig.5-19.

2. Press POWER UP/RESTART on the control keypad, shown in Fig.5-17, which moves the spindle to its machine zero reference location. Fig.5-20 shows machine zero on a mill.

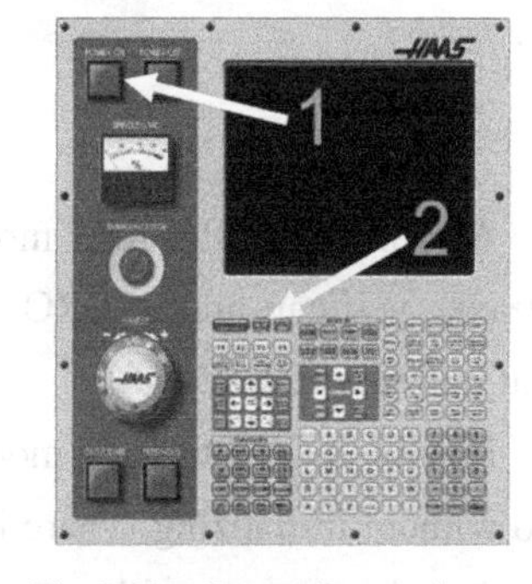

Fig.5-18 Steps of powering up

Fig.5-19 POWER ON and POWER OFF

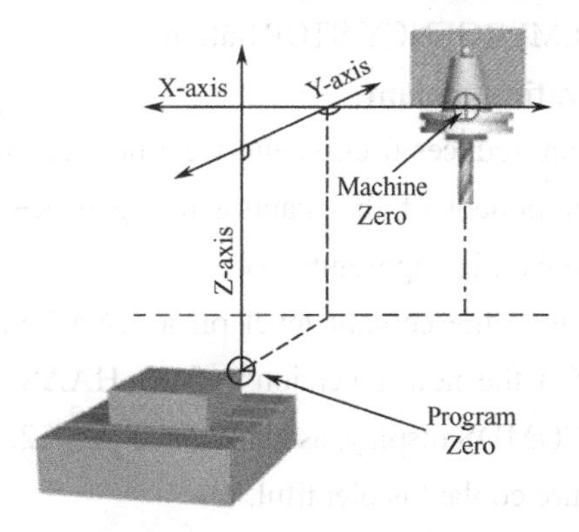

Fig.5-20 Location of machine zero

Machine zero, which is fixed by the manufacturer, is the farthest distance along the positive machine axes. The machine must find its fixed machine reference before it can perform any operations[13].

Because it is such an important position, most machines require that you send the machine to its reference position during the machine's power up procedure.

Remember that the main breaker at the rear of the machine must be switched on before the POWER ON button will turn on the mill. Any power interruption will turn the mill off, and the POWER ON button must be used to turn the power on again.

Zero return

In addition to the POWER UP/RESTART key, there are other ways to get the spindle to the machine zero position.

After the machine is powered up, it will not know its home position until it has been returned to machine zero by the POWER UP/RESTART key. You may also get the spindle to machine zero by pressing the ZERO RET mode key, then pressing the AUTO ALL AXES key. These keys are shown in Fig.5-21. This will home the spindle and initialize all axes to machine zero.

Fig.5-21　ZERO RET mode keys

Also in the ZERO RET mode, if you want to return the spindle to machine zero and initialize one axis, you can press ZERO SINGLE AXIS. This will initialize an axis that is specified in the input buffer with a letter.

Finally, in ZERO RET mode, the HOME G28 key may be used to rapid all axes to machine zero without initializing. There is no warning to alert you of any possible collision, so special care must be taken to ensure the machine has a clear path home[14]. If you sense an impending collision, press the EMERGENCY STOP button.

Activating coolant

Coolant reduces friction and wear between the cutting tool and the workpiece. Since coolant is a vital component of the manufacturing process, checking coolant levels and knowing how to activate coolant is important.

To check the coolant level on a HAAS mill with an older control, you may simply use a dipstick. On the newer versions of the HAAS control, you can view a coolant sensor from the CURNT COMDS display, as shown in Fig.5-22. Regardless of your control version, it is important to make sure coolant is plentiful.

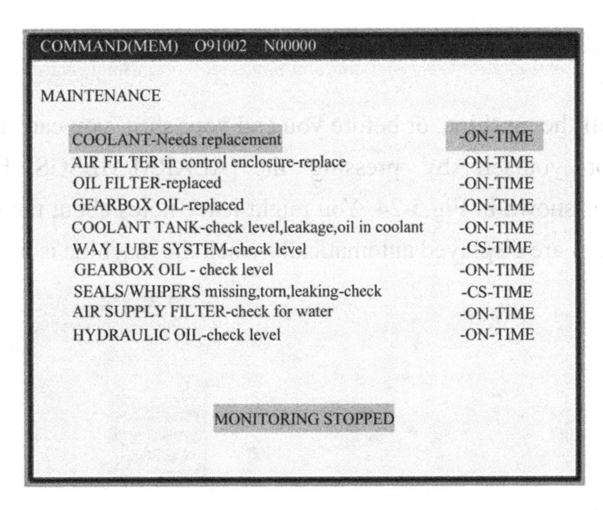

Fig.5-22 Coolant sensor from CURNT COMDS display

The Settings display lets you control the use of coolant during operation. Press the SETNG/GRAPH key to view the list of settings. Setting 32 controls the coolant pump:

- In the "Normal" setting, M08 and M88 coolant commands are executed as programmed.
- In the "Ignore" setting, an M08 or M88 command in the program will not turn on the coolant, but it can be turned on manually using the COOLNT key.
- In the "Off" setting, the coolant cannot be turned on at all, and the control will give an alarm when it reads an M08 or M88 command in a program.

The coolant pump can be turned on or off manually while a program is running by pressing the COOLNT button, which is shown in Fig.5-23. This will override what the program is doing until another M08 or M09 coolant command is executed.

Remember also that in the column to the right of the jog keys are buttons that control the coolant for the machine, as shown in Fig.5-15. If coolant spigots are enabled, CLNT UP positions the coolant stream direction one position higher, the CLNT DOWN positions the coolant stream direction one position lower, and AUX CLNT turns on the Through the Spindle Coolant system if in MDI mode. Pressing AUX CLNT twice shuts off the system.

Activating the chip conveyor

Machining operations create chips. Chips are the metal by-products of metal cutting. Removing chips from the cutting area is important. If you do not remove chips, they may interfere with the cutting capabilities of the machine. You can remove chips by turning on the chip conveyor. The conveyor carries chips to a cart, which is used to move the chips to the shop's central recycling area.

The jog keys that control chip movement are shown in Fig.5-15. The chip conveyor can be turned on or off when a program is running, either manually by using the jog keys or in the program using M codes. CHIP FWD turns the chip auger in a direction that removes chips, CHIP STOP stops chip auger movement, and CHIP REV turns the chip auger in the reverse direction.

M codes are codes that signal miscellaneous commands in the machine control. The M code equivalent to CHIP FWD is M31, CHIP REV is M32, and CHIP STOP is M33. On newer HAAS controls, M32 is no longer an active code for reversing the chip conveyor.

Leaving messages

Before you shut off the machine, or before you end your shift, you can enter a message for the next operator or for yourself by pressing the ALARM/MESGS button twice[15]. The ALARM/MESGS key is shown in Fig.5-24. You might leave notes about the status or maintenance of the machine. Messages are displayed automatically when the machine is powered up.

Fig.5-23 Coolant button

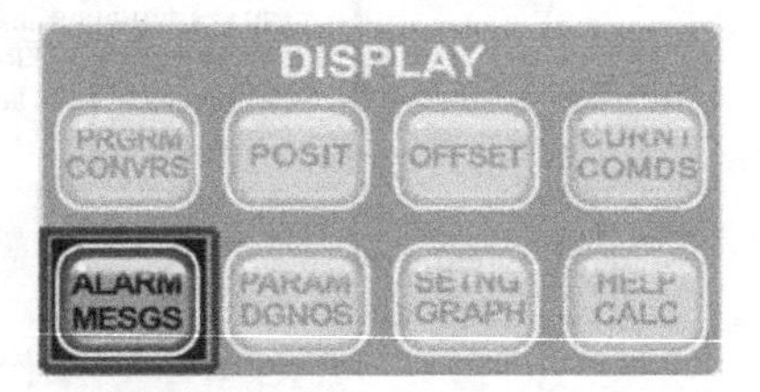

Fig.5-24 ALARM/MESGS key

Occasionally, the machine will need to tell you about the status of its general operation. If you press the ALARM/MESGS key once, the machine will display all current alarms with a number and description. You may remove each alarm one at a time with the RESET key. By pressing either the left or right cursor keys, you can display an Alarm History of the last 100 alarms that have been recognized by the control. The history will include the date and time of the alarm.

Powering down

There are a few ways to power down the HAAS mill. The most common method is to press the red POWER OFF button to remove power from the machine instantly. POWER OFF should not be used while the machine is executing a program. There are also two settings in the SETNG display that will power down the machine under certain conditions:

- The Auto Power Off Timer, or Setting 1, is a numeric setting that indicates how many minutes of idle time must lapse before the machine is automatically turned off. This will not occur while a program is running or while you are pressing keys.
- Power Off At M30, or Setting 2, is a setting that, when set to ON, will power down a machine when an M30 code ends a program.

Both of the above settings, when enabled, will give you 15 seconds before the control shuts off the machine. If you determine that the machine should stay on, pressing any key will interrupt the timer's countdown. Also, these settings will automatically place the tool listed in Setting 81, Tool at Power Down, in the spindle first.

Other conditions may cause the machine to power down. A sustained overvoltage, showing Alarm 176, or overheating condition, showing Alarm 177, will start auto-shutdown. Each condition must last more than 4.5 minutes before the machine shuts down.

Summary

The HAAS control is one of the most common types of CNC controls in the shop today. The HAAS mill control panel has three distinct regions. First, manual controls of the HAAS mill, such as the HANDLE, EMERGENCY STOP, and FEED HOLD key, function much like the manual controls of other machines. Second, the display screen shows all relevant information needed during the machining process. Third, the keypad allows the operator to enter machine commands.

The HAAS mill control keypad has nine areas, each with keys grouped together by similar function and purpose. The control keypad consists of display, operation mode, numeric, alpha, override, cursor arrow, jog, function, and reset keys.

A basic understanding of the control panel and keypad allows the operator to perform a variety of basic tasks, including powering up and powering down, checking and activating coolant, sending the machine to machine zero, activating the chip conveyor, and leaving messages.

TECHNICAL WORDS

specific	[spi'sifik]	*a.& n.*	特定的；细节
keypad	['kiːpæd]	*n.*	键盘
handle	['hænd(ə)l]	*n.*	手轮
override	[ˌəuvə'raid]	*n. & v.*	修正，调整，修调
collide	[kə'laid]	*v.*	碰撞
fixture	['fikstʃə]	*n.*	夹具
capital	['kæpit(ə)l]	*n.*	大写字母；资本
alphabet	['ælfəbet]	*n.*	字母表
cursor	['kəːsə(r)]	*n.*	光标，指针
backspace	['bækspeis]	*n.& v.*	退格
character	['kæriktə(r)]	*n.*	字符；特性
highlight	['hailait]	*v.& n.*	加亮；加亮区
parameter	[pə'ræmitə(r)]	*n.*	参数
diagnostics	[ˌdaiəg'nɔstiks]	*n.*	诊断
auger	['ɔːgə]	*n.*	螺丝钻
void	[vɔid]	*v.*	使……无效
warranty	['wɔrənti]	*n.*	保修（期）
deactivate	[diː'æktiveit]	*v.*	撤销
visually	['vizjuəli]	*ad.*	在视觉上地
tangent	['tændʒənt]	*a.& n.*	相切的；切线
decimal	['desiməl]	*n.*	十进制；小数（点）

alter	[ˈɔːltə]	*v.*	改变，替换
jog	[dʒɔg]	*n.& v.*	点动，手动
initialize	[iˈniʃəlaiz]	*v.*	初始化
friction	[ˈfrikʃən]	*n.*	摩擦
wear	[wɛə]	*n.*	磨损
dipstick	[ˈdipstik]	*n.*	量油计
spigot	[ˈspigət]	*n.*	龙头；喷嘴
shift	[ʃift]	*v.& n.*	轮班；移动
maintenance	[ˈmeintinəns]	*n.*	维护，保养
countdown	[ˈkauntˌdaun]	*n.*	倒计时

PHRASES

control panel		控制面板
lower case letter		小写字母
floppy drive		软盘驱动器
hard drive		硬盘驱动器
on the fly		实时，即时，快速，飞速
machine zero		机床零点
in between		在中间
program block		程序段
input buffer		输入缓冲器
block delete		程序段删除，跳步
macro statement		宏程序语句
work coordinate offset		工件坐标偏置值
through the spindle coolant	(TSC)	主轴中心孔冷却
tool changer		换刀装置
tool release		松刀
part zero	(work zero)	工件零点
power up		上电，通电，开机
main breaker		主（回路）断路器
home position		参考点位置（在铣床上就是机床零点）
home the spindle	(home the machine)	对机床回参考点
chip conveyor		排屑装置
chip auger		螺旋排屑器

negative number	负数
shut down	关机，关断

NOTES

1. While other CNC mills may have machines and controls from different manufacturers, every HAAS mill comes with a HAAS control. 尽管别的数控铣床会采用不同制造厂家的机床本体和数控系统，可每一台 HAAS 铣床都配置 HAAS 数控系统。

2. The display screen shows you the program, axis locations, and other relevant information throughout the machining process. 显示屏显示程序、轴的位置，以及整个加工过程中其他相关的信息。

3. Manual controls such as **HANDLE**, EMERGENCY STOP, CYCLE START, and FEED HOLD **function** much like the controls on other machines. 手轮、紧停按钮、循环启动按钮和进给保持按钮这些手动控制部件的功能很像其他机床上的手动控制部件。*这里 function 是动词，作为谓语，翻译时转换为名词，译成"……的功能"。HANDLE 也称为 electronic handwheel（电子手轮），用手摇动时产生电脉冲使轴运动。*

4. HAAS **advise**s that you not change parameters unless you know exactly what needs to be changed and why, and that you have made all the correct inquiries within your shop and with HAAS service personnel. If a parameter is changed without proper assistance, you may void the warranty of the machine. HAAS 公司建议用户在无法确定需要改什么和为什么要改，以及没有正确咨询车间和 HAAS 服务人员之前，不要随意更改参数。如果参数被随意更改，则用户可能无法得到机床的保修服务。*advise 为主句谓语时，从句须用虚拟语气，使用 should+动词原形或动词原形。第二个 that 为 unless 从句的并列成分。*

5. Pressing SETNG/GRAPH once shows settings, which are machine control functions you may need to activate, deactivate, or change to suit specific conditions. 按一下 SETNG/GRAPH 显示设定页面，用于设定机床控制功能，用户可能需要激活、撤销或更改这些机床控制功能，以符合具体的工况。

6. Pressing SETNG/GRAPH twice allows you to run a part program visually without risking tool or machine damage due to programming errors. 按两下 SETNG/GRAPH，用户可以从屏幕上看到零件程序的运行情况，从而避免由于编程错误造成刀具和机床损坏的危险。

7. The keys in the same row as the pressed mode key are available after a mode has been selected. 选择某一操作方式后，与方式键在同一行的其他键就生效了。

8. HANDLE JOG mode allows you to move the spindle along its axes with either the HANDLE or JOG buttons. HANDLE JOG 方式允许操作员用手轮或手动按钮移动各坐标轴。

9. By using overrides, you can adjust **these variables "on the fly"** while the program is being executed. 通过使用修调键，用户可以在程序运行时调整这些变量。*这里，"these variables"（这些变量）指进给速度、主轴转速、快速移动速度。大部分数控系统具有"on the fly"修调功能。*

10. When you press the JOG LOCK key before one of the above keys, the axis is moved in a continuous motion without the need to hold the axis key **depressed**. 在操作上述键之前按下 JOG LOCK 键，则不用一直按住轴进给键，轴也能连续运动。*depressed 为宾语补足语。*

11. TOOL OFSET MESUR sets the present Z-location for the tool length offsets in the offset page during part setup. TOOL OFSET MESUR 用于零件装夹时在偏置页面中（自动）设定当前 Z 轴刀具长度偏置值。

12. The key initiates a user prompt screen to **assist** the operator **in** recovering from a tool changer crash. 这个键调出一个用户提示页面，帮助操作员进行换刀碰撞故障恢复。

13. Machine zero, which is fixed by the manufacturer, is the farthest distance along the positive machine axes. The machine must find its fixed machine reference before it can perform any operations. 机床零点由机床制造厂设定并且固定不变，它是机床坐标轴正向的极限点。在机床进行任何操作之前，必须先找到其固定的机床参考点。

14. Finally, in ZERO RET mode, the HOME G28 key may be used to rapid all axes to machine zero **without initializing**. There is no warning to alert you of any possible collision, so special care must be taken to ensure the machine has a clear path home. 最后，ZERO RET 方式下，HOME G28 键可用于在不进行初始化的条件下，使机床各轴快速返回机床零点。由于没有警告信号提醒用户可能发生碰撞，因此必须特别小心，以保证机床回参考点时不会与其他物体发生干涉。*这里"不进行初始化"是指没有手动回参考点时寻找编码器零标志脉冲的过程。*

15. Before you shut off the machine, or before you end your shift, you can enter a message for the next operator or for yourself by pressing the ALARM/MESGS button twice. 关机或换班之前，按两下 ALARM/MESGS 按钮，可以给下一位操作员或给操作员自己输入信息。

PRACTICE

Task 1 Translate the following words or phrases into English.

1. 操作方式
2. 倍率
3. 机床零点
4. 工件零点
5. 开机
6. 关机
7. 回零
8. 紧停
9. 工件坐标偏置
10. 刀具偏置
11. 进给保持
12. 执行程序(*v.*)
13. 新建程序(*v.*)
14. 零件装夹
15. 程序段
16. 排屑装置

Task 2 Write the correct form of each word as requested.

implement ____________ (*n.*)　　installation ________________(*v.*)

operation ____________ (*v.*)　　needless ________________(*syno.*)

flexible ____________ (*n.*)　　minimum ________________(*anto.*)

reduce ____________ (*n.*)　　form ________________(*v.*)

advantage ____________ (*anto.*)　　automatic ________________(*v.*)

computer ____________ (*v.*)　　coordination ________________(*v.*)

clockwise	______________	(*anto.*)	increase	______________	(*anto.*)
rigidly	______________	(*n.*)	transmit	______________	(*n.*)
increment	______________	(*a.*)	top	______________	(*anto.*)
collision	______________	(*v.*)	activate	______________	(*anto.*)

Task 3 Fill in the brackets with words that have similar meaning to the underlined words, changing their forms if necessary.

1. (   ) The overrides give you the ability to <u>alter</u> the programmed feed and speed, spindle direction, and rapid traverse motion.

2. (   ) MDI/DNC allows you to enter and <u>execute</u> program commands without creating a stored program or execute a program from another location.

3. (   ) After the machine is powered up, it will not know its <u>home</u> position until it has been returned to machine zero by the POWER UP/RESTART key.

4. (   ) In MEM mode, the current program will start when you <u>hit</u> CYCLE START.

5. (   ) There are a few ways to <u>power down</u> the HAAS mill.

6. (   ) Pressing the COOLNT button will <u>override</u> what the program is doing until another M08 or M09 coolant command is executed.

7. (   ) The most common <u>method</u> is to press the red POWER OFF button to remove power from the machine instantly.

8. (   ) The top row of the display keys provides <u>essential</u> information needed to successfully run a program.

9. (   ) The display screen shows you the program, axis <u>locations</u>, and other relevant information throughout the machining process.

10. (   ) Pressing SETNG/GRAPH twice allows you to run a part program visually without risking tool or machine damage due to programming <u>errors</u>.

11. (   ) The keys in the same row as the pressed mode key are available after a mode has been <u>selected</u>.

12. (   ) 5% RAPID, 25% RAPID, and 50% RAPID allow you to reduce the speed of rapid <u>travel</u> by a specific percentage.

13. (   ) 100% RAPID tells the machine to rapid <u>traverse</u> at its maximum rate.

14. (   ) 100% RAPID tells the machine to rapid traverse at its maximum <u>rate</u>.

15. (   ) Function keys perform <u>a variety of</u> different tasks.

16. (   ) Since coolant is a <u>vital</u> component of the manufacturing process, checking coolant levels and knowing how to activate coolant is important.

17. (   ) Since coolant is a vital component of the <u>manufacturing</u> process, checking coolant levels and knowing how to activate coolant is important.

18. (   ) On the newer versions of the HAAS control, you can <u>view</u> a coolant sensor from the CURNT COMDS display.

19. (　　　　　　　) The coolant pump can be turned on or off manually while a program is running by pressing the COOLNT button.

20. (　　　　　　　) Machining operations create chips.

Task 4　Decide whether the following statements are true (T) or false (F)according to the text.

1. (　　) Turning the handle can override the feed rate or the spindle speed.
2. (　　) Machine zero is set by the programmer or the machine operator.
3. (　　) Machine parameters can be changed by the user, if needed.
4. (　　) There are several ways to get the spindle to machine zero.
5. (　　) After the machine is powered up, it will know its home position immediately.
6. (　　) Settings can be changed by the operator, if needed.
7. (　　) You should push the RESET button immediately if a tool is about to collide with a part or fixture.
8. (　　) When the tool release button is used, the tool can be removed from the spindle. This can be operated in any mode.
9. (　　) For the sake of safety, the tool release button must be held for one-half second before the tool is released.
10. (　　) When the tool release button is released, the tool will be clamped in the spindle immediately.

Task 5　You are required to find the English explanations for the list of terms. Then you should fill the brackets with the corresponding letters.

A. override	G. cycle stop
B. handle mode	H. control panel
C. spindle speed	I. edit mode
D. cycle start	J. part program
E. MDI mode	K. emergency stop
F. offset	L. CNC

Example:　(**L**) *A form of programmable automation in which the machine tool is controlled by a program in computer memory.*

1. (　　) The control that automatically shuts down all machine functions.
2. (　　) The group of controls on a CNC machine that run, store, and edit the commands of a part program and other coordinate information.
3. (　　) The mode that allows for the manual operation of tool movement via the handwheel.
4. (　　) A control that adjusts a programmed speed or feed rate by a certain percentage during operation.
5. (　　) The control used to pause a program.
6. (　　) The mode that allows an operator to manually enter and execute blocks of programming code at the control panel.

7. (　　　) A series of numerical instructions used by a CNC machine to perform the necessary sequence of operations to machine a specific workpiece.

8. (　　　) The control used to begin a program or continue a program that has been previously stopped.

9. (　　　) The mode that allows an operator to make changes to a part program and store those changes.

10. (　　　) The number of revolutions that the spindle makes in one minute of operation.

Task 6　Select an answer from the four (or three) choices.

1. Powering down the HAAS mill occurs when you:

(A) Set the machine to Setting 81: Tool at Power Down

(B) Allow the machine to experience a prolonged undervoltage situation

(C) Press POWER OFF or activate Setting 1: Auto Power Off Timer

(D) Press POWER OFF or activate Parameter 80: Power Off at G88

Your answer: ________

2. On the newer versions of the HAAS control, you can view a coolant sensor from the:

(A) CURNT COMDS display

(B) ALARM/MESGS display

(C) MDI/DNC mode key

(D) LIST PROG mode key

Your answer: ________

3. The HAAS CNC control panel:

(A) Relies on the FEED HOLD button to move machine components along selected axes

(B) Includes the display screen, spindle load meter, and emergency stop button

(C) Uses an emergency stop button for the routine powering down of the machine

Your answer: ________

4. Which is a true statement about the operation mode keys?

(A) ZERO RET allows you to move the spindle with the HANDLE or jog keys

(B) MDI/DNC allows you to enter and execute program commands without creating a stored program or execute a program from another location

(C) MEM allows manual editing changes or the creation of a new program

(D) HANDLE JOG searches for machine zero or rapids home automatically

Your answer: ________

5. Which is a true statement of the function keys?

(A) PART ZERO SET is used to select the next tool during part setup

(B) TOOL OFSET MESUR is used to enter present Z-axis location for the tool length offsets in the offset page during part setup

(C) NEXT TOOL is used to toggle between the X-axis and Z-axis jog mode during setup

Your answer: ________

6. Which is a true statement about the display keys?

(A) POSIT shows the list of machine parameters

(B) OFSET shows modal program values and the position during runtime

(C) PRGRM/CONVRS shows the program blocks of the current part program

Your answer: __________

7. On the diagram (see Fig.5-25), which region of the control keypad represents the location of the alpha and numeric keys?

(A) A　　(B) B　　(C) C　　(D) D

Your answer: __________

8. On the diagram (see Fig.5-26), which group of keys moves the spindle along the X-, Y-, and Z-axis?

(A) A　　(B) B　　(C) C

Your answer: __________

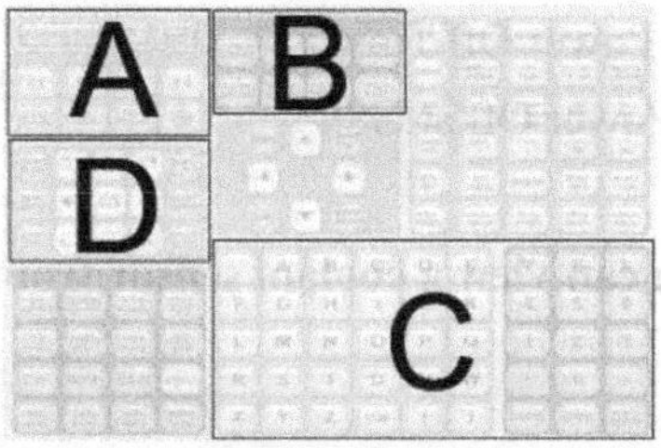

Fig.5-25　Regions of the control keypad

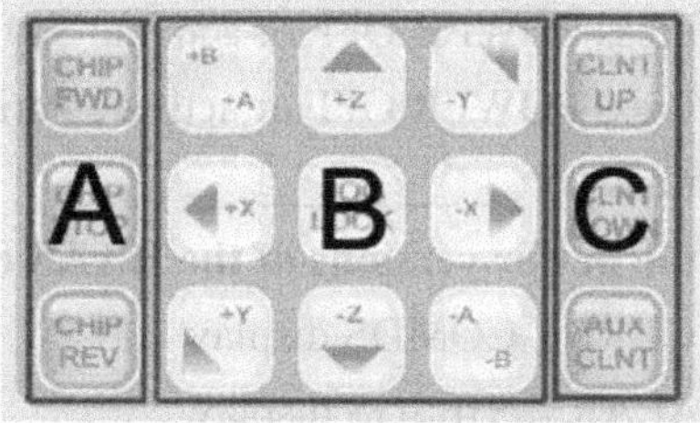

Fig.5-26　Groups of keys

9. The 15 keys located in the lower left corner of the HAAS control keypad (see Fig.5-27) are:

(A) Jog keys

(B) Operation mode keys

(C) Overrides

(D) Display keys

Fig.5-27　Control keypad

Your answer: __________

10. Which is a true statement about the display keys?

(A) SETNG/GRAPH shows machine functions that are activated or deactivated

(B) ALARM/MESGS shows the list of machine parameters

(C) HELP/CALC shows five pages of machine axis positions of the spindle

(D) PARAM/DGNOS shows the program blocks of the current part program

Your answer: __________

11. Compared to other mills, HAAS CNC mills:

(A) Only come with HAAS controls

(B) Have more than one control panel

(C) Use CNC controls from other manufacturers

Your answer: __________

12. Which causes the chip conveyor to remove chips?

(A) Pressing CHP FWD or programming an M31 code

(B) Pressing TS RAPID or programming an M33 code

(C) Pressing FWD or programming an M08 code

Your answer: __________

13. Before the end of a shift, or before machine shut down, an operator may leave a message to another operator by pressing:

(A) ALARM/MESGS once

(B) ALARM/MESGS twice

(C) PRGRM/CONVRS once

(D) PRGRM/CONVRS twice

Your answer: __________

14. In what mode does pressing AUTO ALL AXES, ZERO SINGL AXIS, or HOME G28 move the spindle of the mill to machine zero?

(A) HANDLE JOG

(B) MDI/DNC

(C) ZERO RET

Your answer: __________

15. Which of the following sequences best characterizes powering up?

(A) Make sure main circuit breaker is on; press POWER ON; press POWER UP/RESTART

(B) Press POWER ON; press EOB; press POWER UP/RESTART

(C) Press POWER UP/RESTART; make sure main circuit breaker is on; press CURNT/COMDS

Your answer: __________

16. Which keys allow an operator to alter programmed feed, speed, spindle direction, and rapid traverse motion?

(A) Jog keys (B) Override keys (C) Cursor keys (D) Function keys

Your answer: __________

17. The red RESET button:

(A) Automatically shuts down all machine functions and is used if a tool is about to collide with a part or fixture.

(B) Stops axis motion but continues spindle movement and is used if you want to check the part or tooling.

(C) Stops all machine motion and places the program pointer at the top of the current program.

Your answer: ________

18. If you want to check the part or tooling during the machining process, press:

(A) The FEED HOLD button.

(B) The EMERGENCY STOP button.

(C) The RESET button.

Your answer: ________

Task 7 Fill in the blanks with the following phrases given in the bank.

emergency stop, control keypad, part program, operation mode, machine zero

1. The ________________ is where you enter commands for the control.
2. Pressing PRGRM/CONVRS shows the current selected ________________ .
3. A(n) ________________tells what tasks the control will perform.
4. The ________________ button automatically shuts down all machine functions.
5. ZERO RET mode searches for ____________________ .

Task 8 Write the English name of each component indicated in Fig.5-28 on the corresponding number.

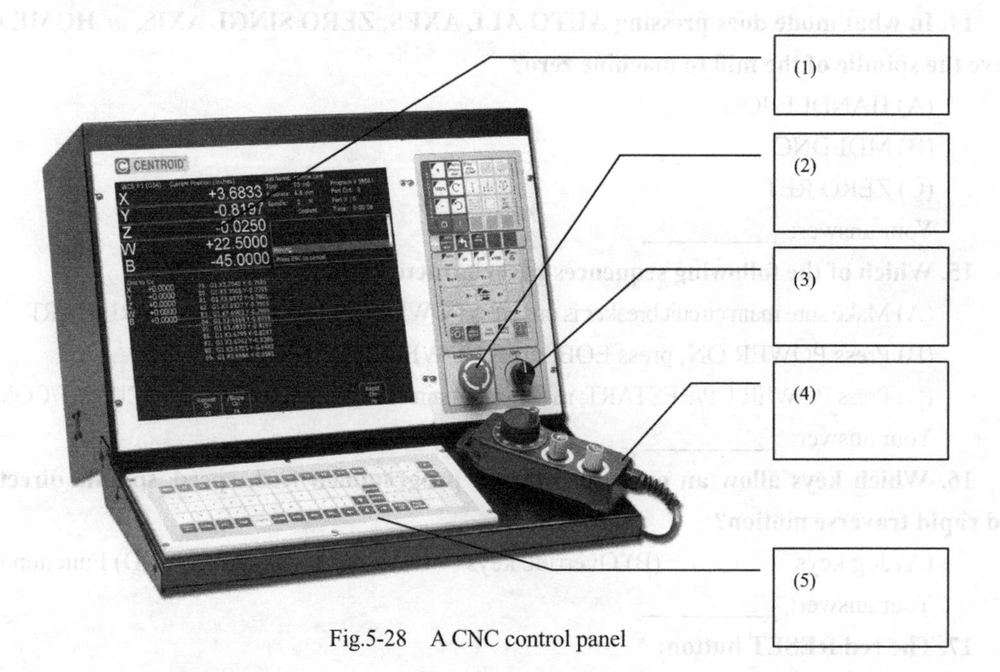

Fig.5-28 A CNC control panel

Task 9 The following information relates to the functions of the control panel. Match Column A ,B, and C.

NO.	Column A	Column B	Column C
1	The display screen	initialize	the programmed feed and speed
2	The SPINDLE LOAD meter	adjust	the axes
3	The HANDLE	indicate	the machine
4	The overrides	show	the power draw
5	The RESET button	move	the information
6	The POWER UP/RESTART button	stop	all machine motion

Task 10 Find the solutions to the following operations. Pick out corresponding sentences or paragraphs from the text.

Operation 1: How to power on the machine? How to power off the machine?

Operation 2: How to reference the machine? How many ways are there?

Operation 3: How to jog the machine and override the feed rate? How many ways are there?

Operation 4: How to program the machine and start machining? How to override the cutting feed rate?

Operation 5: How to start/stop the spindle? How many ways are there?

Operation 6: How to override the spindle speed on the fly? How many ways are there?

Task 11 Find the missing words for the following passage and then read it aloud.

There are six operation modes in the HAAS control:

- ________ mode allows manual editing changes in a program or creation of a new program. You may INSERT, ALTER, DELETE, or UNDO any program.
- ________ mode runs part programs from the control's memory. Pressing MEM shows the current program which will start when you hit CYCLE START.
- ________ mode places the machine in Manual Data Input mode. MDI lets you enter and execute program data without disturbing the stored programs. Pressing MDI/DNC twice activates Direct Numerical Control (DNC) if Setting 55 is on. DNC allows you to execute programs sent from a floppy drive or a computer hard drive.
- ________ mode allows you to move the spindle along its axes with either the HANDLE or JOG buttons.
- ________ mode searches for machine zero or rapid returns to machine zero automatically.
- ________ mode lists programs and allows you to select, send, receive, and delete programs. To create a new program, you must be in the PRGRM/CONVRS display and LIST PROG mode. Enter **Onnnnn** from the keyboard and press SELECT PROG button. Press EDIT to show the new program.

Part B Listening

Task 1 Listen to the five statements twice and write them down.

1. --
2. --
3. --
4. --
5. --

Task 2 The following video is about HAAS factory tour. Watch it first, then listen to it twice and answer the following questions according to what you hear. Pay close attention to the figures mentioned in the video.

Question 1: Where is the factory located? What's the full name of the company?

HAAS

Question 2: How many engineers design HAAS machines?

Question 3: How many CNC chip-making machines are there operating in HAAS machine shop?

Question 4：How many machines and pallets does the simplest HAAS FMS hold?

Question 5：How many machines, pallets and tools does the largest HAAS FMS hold?

Task 3 Watch the above video once more, then listen to it twice and fill in the following blanks with what you hear and see.

While HAAS machines are conceived（设想）as ideas and (1)___________drawings, they are born of hundreds of individual (2)__________, most of which are manufactured in house on state-of-the-art（最先进的）CNC (3)__________ equipment. At the heart of the facility（设施）, you'll find the manufacturing center of HAAS AUTOMATION—the machine shop. There are more than 270 CNC (4)______________ machines in operation here, machines that consume upwards of （……以上）136 million pounds of raw material（原材料）per year. How do we maintain the HAAS level of (5)__________ in such a high output production environment? We put nearly 200 HAAS machines to work for us right in our own machine (6)__________. This allows us to observe firsthand the day-to-day (7)___________ of our machine tools, many of which employ some of the most advanced manufacturing technologies in use today, such as robotic machining (8)__________ that run unattended（无人照看的）. We also employ flexible（柔性的）(9)___________ systems that range from a single machine with 6 (10)_______ to our largest custom-configured system with 5 machines, 143 pallets, and 1650 tools.

Task 4 The following video is about axis jogging. Watch it and choose the best answer to each of the following questions.

1. What is jogging?

(A) Manual operation of tool movement via the handwheel.

(B) Automatic operation of tool movement via the program.

(C) Manual operation of tool movement via the buttons.

Your answer: _________

2. When holding down the X+ jog key, the table:

(A) is moving to the right.

(B) is moving to the left.

(C) is moving towards the operator.

Your answer: _________

3. When holding down the Y+ jog key, the table:

(A) is moving to the right.

(B) is moving away from the operator.

(C) is moving towards the operator.

Your answer: _________

4. **When holding down the Z- jog key, the tool:**

(A) is moving up.

(B) is moving down.

(C) is moving left.

Your answer: ________

5. **When holding down the Y- jog key, the tool:**

(A) is moving towards the operator.

(B) is moving down.

(C) is moving left.

Your answer: ________

Part C Speaking

Task 1 Watch the slides and give the English name or description for each of the slides. Take notes.

1. .... 6. ....
2. .... 7. ....
3. .... 8. ....
4. .... 9. ....
5. .... 10. ....

Task 2 Work in pairs. Take turns answering the following questions. Take notes.

1. How many parts does the HAAS control panel consist of?
2. What do the manual controls include?
3. Does the FEED HOLD button stop all the machine motion?
4. What purpose does the spindle load meter serve?
5. If you'd like to monitor both the position and the program, which key you should press?
6. Can you tell the full name of the following words: MDI, DNC, EOB?
7. How to start a program from the control's memory?
8. How many operation modes are there?
9. What purpose do the overrides serve?
10. What is the machine zero?

Task 3 Watch the video clip about axis jogging with the loudspeaker mute and tell the class how to jog the axis on the CNC machine just as an instructor. Take notes.

....

....

....

....

Task 4 ***Shop floor practice. Operate the HAAS machining center in the machine shop and tell your partner how to perform the basic operations with HAAS control. Select one operation, say, how to reference (home) the CNC machine, and apply the job instruction method learned in LS3.***

Job Breakdown Sheet

Task 5 Work in groups. Create a situation about receiving a customer and role-play the situation before the class. One acts as the guide, some others as the visitors.

WORKING SITUATION

Some foreign guests visit your company. You are required to show the visitors around the machine shop. Are you ready for **customer reception**?

【引导文】接待客户参观用语 Reception

A: Welcome to our factory! 欢迎到我们的工厂来！

B: Yes. The surrounding is quite good. 嗯，环境很好。

A: Come this way please. 跟我来。

B: Thank you. 谢谢。

A: This way. 这边请。

B: After you. 您先。

A: Would you like to go through our factory some time?什么时候来看看我们的工厂吧？

B: That's a good idea. 好啊。

A: I can set up a tour next week. 我可以安排在下个礼拜参观。

B: Just let me know which day. 决定好哪一天就告诉我。

A: The tour should last about one hour and a half. 这次参观大概需要一个半小时。

B: I'm really looking forward to this. 我期待这次参观已久了。

A: We can start over here. 我们可以从这里开始。

B: I'll just follow you. 我跟着你就是。

A: Keep you outside the yellow line. Please stop me if you have any question. 在黄线外面走。有任何问题，请随时叫我停下来。

B: I will. 好的。

A: Duck your head as you go through the door there. 经过那儿的门时，请将头放低。

B: Thank you. 谢谢。

A: You'll have to wear this hard hat for the tour. 参观时必须戴上这安全帽。

B: This one seems a little small for me. 这顶我戴好像小了一点。

A: Here, try this one. 喏，试试这一顶。

B: That's better. 好多了。

A: That's the end of the tour. 参观就此结束了。

B: It was a great help to me. 真是获益良多。

A: Just let me know if you want to bring anyone else. 如果你要带别人来，请随时通知我。

B: I'd like to have my boss go through the plant some day. 我真想叫我老板哪天过来看看。

A: Welcome to our showroom. 欢迎参观我们的展示室。

B: Thank you, I'm glad to be here. 谢谢，我很高兴到这里来。

A: Is there anything I can show you? 有什么要我拿给你们看的吗？

B: I think I'd like to just look around. 哦，我只是看看而已。

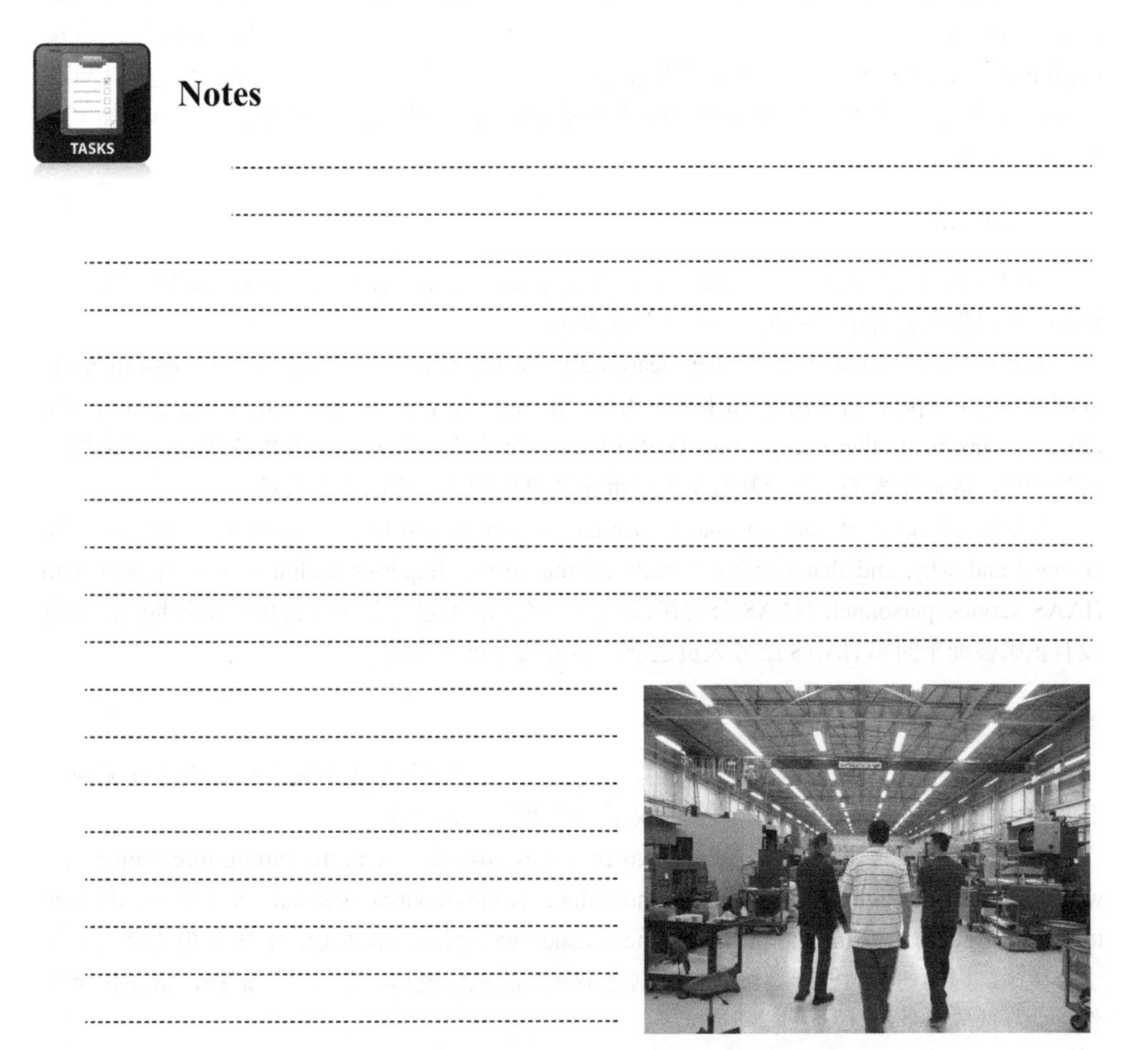

Notes

Part D Grammar and Translation

长句的译法

对科技英语中复杂的长句有下列处理方法。

1. 顺译法

长句叙述层次与汉语相近时，可按英语原文顺序依次译出。

They use their knowledge of the working properties of metals and their skill with machine tools to plan and carry out the operations needed to make machined products that meet precise specifications. (Ref. LS 1) 他们运用其金属材料特性方面的知识和机床方面的技能进行工艺规划和加工，制造满足精度要求的机加工产品。

Since this form of CNC machine can perform multiple operations in a single program (as many CNC machines can), the beginner should also know the basics of how to process workpieces machined by turning so a sequence of machining operations can be developed for workpieces to be machined. (Ref. LS 4) 由于这种数控机床能在一个程序中完成多种加工（很多数控机床都能够如此），初学者同时应该了解如何加工车削类零件的基础知识，这样才能编制出待加工零件的加工工序。

2. 逆译法

有时英语长句的展开层次与汉语表达方式相反，这时就需要逆着原文的顺序译出。英语的表达习惯是先说出主要的，然后才说次要的。

Therefore, the number of workers learning to be machinists is expected to be less than the number of job openings arising each year from the need to replace experienced machinists who retire or transfer to other occupations. (Ref. LS 1) 每年都有一些有经验的机械师退休或跳槽到其他职业，因此带来的工作空缺数量大于准备从事机械工工作的工人数量。

HAAS advises that you not change parameters unless you know exactly what needs to be changed and why, and that you have made all the correct inquiries within your shop and with HAAS service personnel. HAAS公司建议用户，在无法确定需要改什么和为什么要改，以及没有正确咨询车间和HAAS服务人员之前，不要随意更改参数。

3. 分译法

有时英语长句中嵌套多重定语从句，或者英语长句中各个主要概念在意义上并无密切联系时，可以拆成独立的短句，再按照汉语习惯重新安排次序。

The headstock is required to be made as robust as possible due to the cutting forces involved, which can distort a lightly built housing, and induce harmonic vibrations that will transfer through to the workpiece, reducing the quality of the finished workpiece. (Ref. LS 2) 床头箱的制造必须十分坚固，因为切削力的影响可能会使制造不坚固的机座变形，并且产生谐振，谐振传至工件，会降低成品件的质量。

Task Read the following text and translate it into Chinese.

The load meter measures the power to the spindle motor. At 100%, the spindle motor can be operated continuously. The 150% level can be sustained（保持） for no more than ten minutes, and at 200% level no more than three minutes. After the specified time, the spindle may begin to slow and even stall（静止）. A 200% load should be reduced to 150% by reducing spindle speed or decreasing the feed rate. Spindle load may increase temporarily（暂时） during speed changes.

Learning Situation 6

Maintenance and troubleshooting

Focus of the situation

The required specifications must be followed in order to keep the machine in good working order and protect the warranty. A pattern for servicing the CNC should be followed in, first, determining the problem's source and, second, solving the problem. [为了使机床保持良好的工作状态，使用户的维护保修权利得到保护，必须遵守操作规定。CNC 维修应遵循的原则是：碰到故障应首先得出正确的判断，然后再加以解决。]

Field work

Search the Internet for information about zero return troubleshooting.

Part A Reading

Maintenance for the horizontal machining center

The following is a list of required regular maintenance for a horizontal machining center as shown in Fig.6-1. Listed are the frequency of service, capacities, and type of fluids required. These required specifications must be followed in order to keep your machine in good working order and protect your warranty[1].

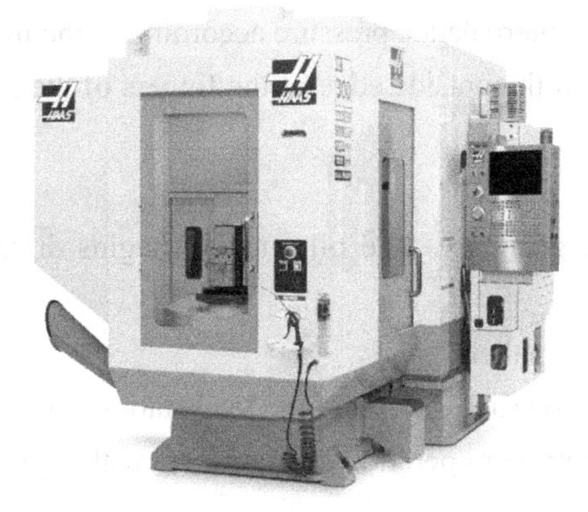

Fig.6-1 Horizontal machining center

Daily

- Top off coolant level every eight hour shift (especially during heavy TSC usage).
- Check way lube lubrication tank level.
- Clean chips from way covers and bottom pan.
- Clean chips from tool changer.
- Wipe spindle taper with a clean cloth rag and apply light oil.

Weekly

- Check for proper operation of auto drain on filter regulator. See Fig.6-2.

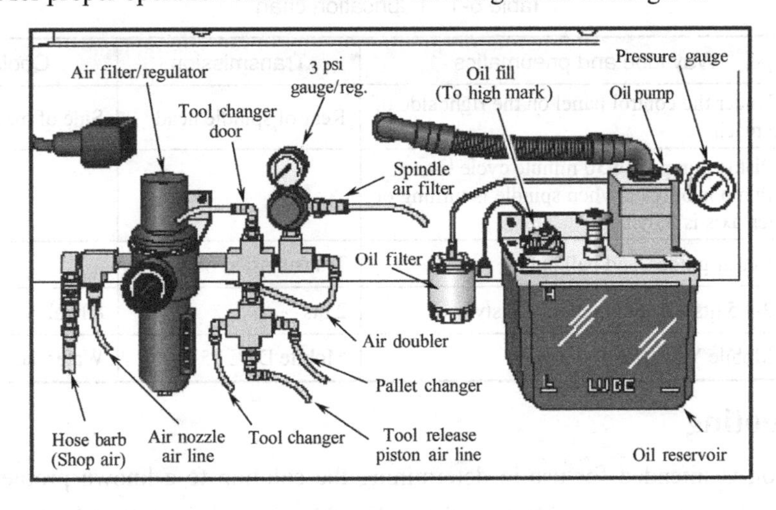

Fig.6-2 Way lube and pneumatics

- On machines with the TSC option, clean the chip basket on the coolant tank.
- Remove the tank cover and remove any sediment inside the tank. Be careful to disconnect the coolant pump from the controller and POWER OFF the control before working on the coolant tank[2]. Do this monthly for machines without the TSC option.
- Check air gauge/regulator for 85 psi.
- For machines with the TSC option, place a dab of grease on the V-flange of tools. Do this monthly for machines without the TSC option.
- Clean exterior surfaces with mild cleaner. Do not use solvents.
- Check the hydraulic counterbalance pressure according to the machine's specifications.
- Place a dab of grease on the outside edge of the fingers of the tool changer and run through all tools[3].

Monthly

- Check oil level in the gear box. Add oil until oil begins dripping from overflow tube at bottom of sump tank.
- Clean pads on bottom of pallets.
- Clean the locating pads on the A-axis and the load station. This requires removing the pallet.
- Inspect way covers for proper operation and lubricate with light oil, if necessary.

Six months

- Replace coolant and thoroughly clean the coolant tank.
- Check all hoses and lubrication lines for cracking[4].

Annually

- Replace the gear box oil. Drain the oil from the gear box, and slowly refill it with 2 quarts of Mobil DTE 25 oil.
- Check the oil filter and clean out residue at the bottom of filter.

See Table 6-1 for the lubrication chart.

Replace the air filter on the control box every 2 years.

Table 6-1 Lubrication chart

System	Way lube and pneumatics	Transmission	Coolant tank
Location	Under the control panel on the right side of the machine	Rear of spindle head	Side of machine
Description	Piston pump with 30 minute cycle time Pump is only on when spindle is turning or when axis is moving		
Lubricates	Linear guides, and ball nuts	Transmission only	
Quantity	2-2.5 qts depending on pump style	2 qts	80 gal
Lubricant	Mobile Vactra#2	Mobile DTE 25	Water based coolant only

Troubleshooting

This section is intended for use in determining the solution to a known problem. Solutions given are intended to give the individual servicing the CNC a pattern to follow in, first, determining the problem's source and, second, solving the problem[5].

Use common sense

Many problems are easily overcome by correctly evaluating the situation. All machine operations are composed of a program, tools, and tooling. You must look at all three before blaming one as the fault area.

If a bored hole is chattering because of an overextended boring bar, don't expect the machine to correct the fault. Don't suspect machine accuracy if the vise bends the part. Don't claim hole mis-positioning if you don't first center-drill the hole.

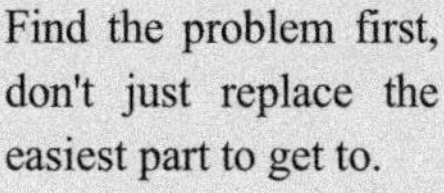

Find the problem first

Many mechanics tear into things before they understand the problem, hoping that it will appear as they go. We know this from the fact that more than half of all warranty returned parts are in good working order. If the spindle doesn't turn, remember that the spindle is connected to the gear box, which is connected to the spindle motor, which is driven by the spindle drive, which is connected to the I/O BOARD, which is driven by the MOCON, which is driven by the processor. The moral here is don't replace the spindle drive if the belt is broken. Find the problem first; don't just replace the easiest part to get to.

Don't tinker with the machine

There are hundreds of parameters, wires, switches, etc., that you can change in the machine. Don't start randomly changing parts and parameters. Remember, there is a good chance that if you change something, you will incorrectly install it or break something else in the process[6]. Consider for a moment changing the processor's board. First, you have to download all parameters, remove a dozen connectors, replace the board, reconnect and reload, and if you make one mistake or bend one tiny pin it WON'T WORK. You always need to consider the risk of accidentally damaging the machine anytime you work on it. It is cheap insurance to double-check a suspect part before physically changing it. The less work you do on the machine the better.

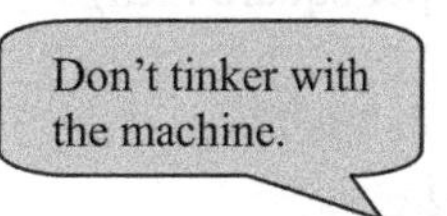

TECHNICAL WORDS

maintain	[mei'tein]	*v.*	保持；维护，保养
troubleshoot	['trʌbl,ʃu:t]	*v.*	故障诊断与维修
service	['sə:vis]	*v.& n.*	服务，维修
fluid	['flu(:)id]	*n.*	液体，切削液
apply	[ə'plai]	*v.*	应用；涂，敷

sediment	['sedimənt]	*n.*	沉淀物
flange	[flændʒ]	*n.*	边缘，法兰
grease	[gri:s]	*n.*	油脂
exterior	[eks'tiəriə]	*a.*	外部的，表面的
hydraulic	[hai'drɔ:lik]	*a.*	液压的，水压的
counterbalance	[ˌkauntə'bæləns]	*v.*& *n.*	平衡
hose	[həuz]	*n.*	软管
residue	['rezidju:]	*n.*	残渣
tooling	['tu:liŋ]	*n.*	（用刀具）加工
moral	['mɔrəl]	*n.*	是非原则

PHRASES

horizontal machining center (HMC)	卧式加工中心
way lube	导轨润滑油
lubrication tank	润滑油箱
way cover	导轨防护罩
spindle taper	主轴锥孔
air filter regulator	空气过滤调压阀
air gauge	气压表
psi(pound per square inch)	气压单位（1 psi=6.89 kPa）
quart(qt)	夸脱，容量单位（1 qt =0.946 L）
gallon(gal)	加仑，容量单位（1 gal =4 qt）
sump tank	废油罐
oil filter	滤油器
common sense	常识
boring bar	镗刀杆
tinker with	胡乱地修补
processor's board	中央处理器板

NOTES

1. These required specifications must be followed in order to keep your machine in good working order and protect your warranty. 为了使机床工作正常，保护保修权利，必须遵守这些必要的规范。

2. Be careful to disconnect the coolant pump from the controller and POWER OFF the control before working on the coolant tank. 注意，处理冷却箱前要关闭冷却液泵，并切断数控系统电源。

3. Place a dab of grease on the outside edge of the fingers of the tool changer and run through

all tools. 涂少量油脂于换刀装置机械手的外边沿，并对全部刀具都用机械手换一遍。

4. Check all hoses and lubrication lines for cracking. 检查所有软管和润滑管路是否破裂。

5. **Solutions given** are intended to give the individual **servicing the CNC** a pattern to follow in, first, determining the problem's source and, second, solving the problem. 以下提出的解决方法用于给数控机床维修人员提供一个可遵循的模式：首先，确定问题的根源，其次，解决问题。*solutions given 中 given 是过去分词作为后置定语；servicing the CNC 修饰 the individual。*

6. There is a good chance that if you change something, you will incorrectly install it or break something else in the process. 如果在维修过程中修改参数、线路等，很可能引起安装错误或破坏其他元件。*There is a good chance that…是很可能……的意思。*

Practice

Task 1 Translate the following phrases into English.

1. 主轴锥孔
2. 导轨防护罩
3. 润滑油箱
4. 故障诊断
5. 维护保养
6. 中心钻
7. 气压表
8. 滤油器

Task 2 Translate the following phrases into Chinese.

1. Horizontal Machining Center
2. in good working order
3. hundreds of parameters, wires, switches
4. spindle drive
5. lubrication tank level
6. hole mis-positioning
7. spindle taper
8. boring bar

Task 3 Fill in the brackets with words that have similar meaning to the underlined words, changing their forms if necessary.

1. ( ) Inspect way covers for <u>proper</u> operation and lubricate with light oil, if necessary.

2. ( ) Many problems are easily <u>overcome</u> by correctly evaluating the situation.

3. ( ) All machine operations <u>are composed of</u> a program, tools, and tooling.

4. ( ) You must look at the program, tools, and tooling before blaming one as the <u>fault</u> area.

5. ( ) Don't suspect machine <u>accuracy</u> if the vise bends the part. Don't claim hole mis-positioning if you don't first center-drill the hole.

6. ( ) Consider for a moment <u>changing</u> the processor's board.

7. ( ) Don't suspect machine accuracy if the <u>vise</u> bends the part.

8. ( ) These required specifications must be <u>followed</u> in order to keep your machine in good working order and protect your warranty.

9. ( ) Listed are the <u>frequency</u> of service, capacities, and type of fluids required.

10. ( ) <u>There is a good chance that</u> if you change something, you will incorrectly

install it or break something else in the process.

Task 4 Decide whether the following statements are true (T) or false (F).

1. ( ) If a bored hole is chattering, something must have been wrong with the machine.
2. ( ) We mustn't start randomly changing parts and parameters.
3. ( ) Don't claim hole mis-positioning if you don't first center-drill the hole.
4. ( ) Top off coolant level every month.
5. ( ) If the spindle doesn't turn, we have to replace the spindle drive.

Task 5 Choose the best answer.

1. How often do we check the way lube lubrication tank level?

(A) Daily. (B) Monthly. (C) Annually.

Your answer: ________

2. How often do we replace the coolant?

(A) Monthly. (B) Six months. (C) Annually.

Your answer: ________

3. How often do we check air gauge for regulated value?

(A) Six months. (B) Monthly. (C) Weekly.

Your answer: ________

4. How often do we clean chips?

(A) Daily. (B) Weekly. (C) Monthly.

Your answer: ________

5. Which of the following are the two main functions of the cutting fluid in machining?

(A) Improve surface finish on the workpiece and wash away chips.

(B) Remove heat from the process and reduce friction at the tool-chip interface.

(C) Reduce forces and power.

Your answer: ________

6. What's the proper sequence of changing the processor's board?

① remove a dozen connectors

② reconnect

③ reload

④ download all parameters

⑤ replace the board

(A) ①→④→③→②→⑤

(B) ④→①→⑤→②→③

(C) ①→②→③→④→⑤

Your answer: ________

7. In addition to reducing friction, what other jobs can lubricants perform when properly used?

(A) Cool the machine and minimize corrosion（腐蚀）.

(B) Ensure that a machine never needs maintaining.

(C) Clean off the various parts of the machine.

Your answer: ________

8. Which is a true statement about the CNC machine lubrication according to the text?

(A) Once the machine is switched on, the oil pump for the way lube is on.

(B) The lubricant for the way lube is different from that for the spindle transmission.

(C) The oil pump cycle time is 60 minutes.

Your answer: ________

Task 6 Match the transitive verbs with the objects in the column on the right.

Transitive verbs	Objects
drill	the program
follow	the problem
perform	the hole
press	the task
execute	the key
solve	the practice

Task 7 Fill in the blanks according to the text, then read the passage aloud.

Many problems are easily ________ by correctly evaluating the situation. All machine operations are composed of ____________________. You must look at all three before blaming one as the fault area.

If a bored hole is________ because of an overextended boring bar, don't expect the ______ to correct the fault. Don't suspect machine ________ if the vise bends the part. Don't claim hole mis-positioning if you don't first ________ the hole.

Part B Listening

Task 1 Listen to the five statements twice and write them down.

1. --

2. --

3. --

4. --

5. --

Task 2 The following video clip is about power setup for the CNC control. Watch it first, then listen to it twice and answer the following questions according to what you hear and see.

Question 1：In which country may the machine be used?

Question 2：What is the voltage between any two phases of the three-phase power source according to the video?

Question 3：What is the voltage between each phase and the ground according to the video?

Question 4：What color is the ground leg?

Question 5：How to check the voltage coming in from your power source?

【语音提示】

Vocabulary		Notes
hook up the power	电源连接	北美相电压为 120V，线电压为 208V
electrical cabinet	电柜	
terminal block	接线端子	
voltmeter	电压表	
hot/ dead	有电/没电	
juice: electricity	电	
three phase/single phase	三相/单相	
plug in	进线插头	

Task 3 The following video clip is about homing the CNC control. Watch it first, then listen to it twice and fill in the blanks according to what you hear and see.

After you power up the machine, what you'll do next is to __________the machine. When the control boots up（启动）, it asks you "Press CYCLE START to send machine to home position" . And it doesn't have a ________ which means digital readout display, because the control doesn't know where it's at. So all you have to do is to press _____________. The homing process is started. It moves one axis at a time, _____-axis first, then _____-axis, then _____-axis.

Part C Speaking

Task 1 Watch the slides and give the English name or description for each of the slides. Take notes.

1. ....
2. ....
3. ....
4. ....
5. ....
6. ....
7. ....
8. ....
9. ....
10. ....

Task 2 Work in pairs. Take turns asking your partner 5 or more questions on machine maintenance frequencies and your partner answers. One question has been given for an example.

1. How often to top off the coolant level?
2. ....
3. ....
4. ....
5. ....

Task 3 Look at the pictures on page 130 and make a list of words or phrases to describe what you see. Then make a dialogue with your partner.

....

....

....

....

....

....

Task 4 Find warnings and cautions labeled on the HAAS machining center. Understand the maintenance and safety measures and tell your partner about them. Make a record by writing them down or taking photos.

....

....

....

....

....

Task 5 Troubleshoot the machine based on the alarm messages on the screen. Make a situational dialogue or conversation, removing the following alarm shown in Fig.6-4.

```
ALARM MESSAGE                      O0917 N00000

 1002    NO ZERO

                                   S     0 T0000
 MDI **** *** *** ALM 15:48:55
(ALARM )( MSG   )(HISTRY)(        )(          )
```

Fig.6-4 An alarm message screen

Task 6 Work in pairs. Jenny is the operator of a horizontal machining center. It appears that the machine can't find its machine zero. Bob, her colleague, helps her solve the problem. You may search information on the Internet with the following keywords and hints.

【Key words】zero return, troubleshooting

【Hints】encoder, deceleration dog, searching velocity

Task 7 Work in pairs. Suppose there is a problem with your machine and you have to contact the service technician for help by telephone. Make a situational dialogue.

WORKING SITUATION

You will probably encounter problems with your machine. You need **troubleshooting by telephone**.

【引导文】电话用语 Telephone calls

● *开始通话:*

Hello! Is that you, Tom? 喂，你是汤姆吗?

Is Daisy there? Daisy 在吗?

Speaking. 我就是。

Who's calling? 你是哪一位?

This is Daisy. 我就是 Daisy。

That's me. 我就是。

- *希望给时间记录：*

Sure, if you can excuse me for just a second. Let me find a piece of paper to write it down.

- *听不清楚，希望对方重复：*

Can you repeat again, please? 能不能请你再重复一次?

(Say) Again, please? 再说一次好吗?

Pardon? 抱歉，请再说一次。

Come again, please? 再说一次好吗?

- *结束通话：*

Nice talking to you. 很高兴跟你通话。

I'll be looking forward to hearing from you. 我期盼你的回音。

- *示例：*

A: Haas Inc. Hi, Mary speaking. 哈斯公司，您好！我是 Mary。

B: Hello, I'd like to speak to Mr. Hunter, please. 您好！我想找 Hunter 先生。

A: May I ask who is calling, please? 请问您是哪位？

B: My name is Herbert Wood of Transcend Manufacturing Company. 我是 Transcend 制造公司的 Herbert Wood.

A: Thank you, Mr. Wood. One moment, please…谢谢，Wood 先生，请稍等……

TASKS

Notes

Part D Grammar and Translation

省译、增译和转译

英汉两种语言，由于表达方式不尽相同，将英语翻译成汉语时有时需要减词，有时需要增词，有时需要转换词性或句子成分，这样才能符合汉语的表达习惯；同时在用英语表达中文意思时，也要注意这种翻译特点，养成以英语思考的习惯，这样才能说出地道的英语，切忌将英语的某词与中文的某词机械地一一对应。

1. 省译

在进行翻译时，原文中的某些词可不必翻译出来，这就是通常说的省译或减词。在不影响意思表达的基础上，一些形式主语、连词、冠词、介词等虚词，一些可有可无或者有了反而觉得累赘的词都可以删减，以符合汉语表达习惯。如：

It is important to recognize **that**, in a turning operation, each cutting pass removes twice the amount of metal indicated by the cross slide feed divisions. (Ref. LS 3) 在外圆车削过程中，每次进刀的金属去除量是中拖板进给刻度指示的两倍，认识到这一点很重要。*it，that 这些虚词可以不译。*

These rigid machine tools remove material from a rotating workpiece via the linear movements of various cutting tools, such as tool **bits** and drill **bits**. (Ref. LS 2) 这些刚性的机床通过各种刀具（如车刀和钻头）相对旋转工件沿直线运动从而去除材料。*这里 bits 表示带刀尖的刀具，车刀和钻头就是这样的刀具，因此可不译。*

You can use the carriage handwheel to **crank** the carriage back to the starting point **by hand**. (Ref. LS 3) 可以使用横拖板手轮将横拖板摇回起点。*这里 crank 就有用手摇的意思，因此 by hand 显得有点累赘，可以省译。*

2. 增译

翻译时也常常需要添词，这样才能使读者明白和易于理解。

（1）抽象名词后加名词使其具体化

翻译某些由动词或形容词派生来的抽象名词时，可根据上下文在其后增添适当的名词，使其更符合汉语习惯。如增译“功能”、“效应”、“现象”、“方法”、“装置”、“变化”等。

A feedback device at the opposite end of the ball screw allows the **control** to confirm that the commanded number of rotations has taken place. (Ref. LS 4) 安装在滚珠丝杠另一端的反馈装置使数控<u>装置</u>能确认，实际上是否转动了被指令的转数。

On machines with the TSC option, clean the chip basket on the coolant tank. 对于带主轴中心孔冷却（TSC）选择<u>功能</u>的机床，清理冷却箱上的切屑收集篮。

（2）不及物动词后增加宾语

英语中有些动词既可及物也可不及物。不及物时，宾语实际上是隐含在动词后面的，译成汉语时往往需要把它们表达出来。

（3）增加表达时态的词

英语动词时态靠词形变化（go, went）或加助动词（will go，have gone）来表达。汉语动词没有词形变化，英译汉时要增加汉语特有的时态助词或表示时间的词，如“曾、了、过、着、将、会”等。此外，为了强调时间概念或时间上的对比，往往也需要增加一些其他的词。

Carbide tipped tools **will** stand speeds in excess of those recommended for high-speed steel tools. (Ref. LS 3) 硬质合金刀具可以承受表中推荐的基于高速钢刀具的转速。

（4）增加反映背景情况的词、解说性的词或句

增加反映背景情况的词、解说性的词以使汉语语意明确，根据上下文增词，符合业内人士的表达习惯。

Know where the **emergency stop** is before operating the lathe. (Ref. LS 3) 操作车床前要知道紧急停止按钮的位置。*这里 emergency stop 即是 emergency stop button，增译“按钮”二字。*

Keep tools overhang as short as possible. (Ref. LS 3) 在不影响操作的情况下，刀具伸出部分越短越好。*增译“在不影响操作的情况下”使语意更加严密。*

The constant temperature workshop consists of large parts machine line, middle and small shell parts machine line, shaft and plate parts machine line, precision machine line, unit assembly line, powder spraying line, final assembly line, automatic solid storehouse, and **precision quality test**. (Ref. LS 7) 恒温车间建有大件加工线、中小壳体类零件加工线、轴类及盘类零件加工线、精密加工线、部件装配线、涂装作业线、总装作业线、全自动立体仓库、精密检测室。*根据句子意思，precision quality test 应增译为精密检测室。*

With Mazak CAMWARE, MAZATROL programs can be generated by a PC. With the Cyber Tool Management of CPC, tool managers can monitor the state of the tools for on line machines and prepare spare tools for replacing worn tools. (Ref. LS 7) 通过 CAMWARE 模块，MAZATROL 加工程序可由计算机生成。通过 CPC 的智能刀具管理模块，刀具管理人员可以监控在线机床的刀具使用状态，准备备用刀具，及时更换超出寿命期的磨损刀具。*根据上下文，增加“模块”二字。*

（5）增添原句中省略的成分

如：英语倒装句中为了避免句子部分内容重复而省略的部分，译成汉语时应把它表达出来。

3. 转译

由于英汉两种语言表达方式不同，语言结构形式差别很大，为了使译文符合汉语表达习惯，除了运用词类转译技巧之外，往往还伴随着句子成分的转换。例如，在某些特定的句型中，英语句子的状态、定语、表语可能转译成汉语句子的谓语等。下面仅列举常见的几种情况，供举一反三、类比推敲。

（1）词类转译

（a）动词转换成名词

It is important to recognize that, in a turning operation, each cutting pass **removes** twice the amount of metal indicated by the cross slide feed divisions. (Ref. LS 3) 在外圆车削过程中，每次

进刀的金属去除量是横向拖板进给刻度指示的两倍，认识到这一点很重要。

Manual controls such as HANDLE, EMERGENCY STOP, CYCLE START, and FEED HOLD **function** much like the controls on other machines. (Ref. LS 5) *见 LS5 之 NOTES 3*。

Rooms with constant temperature and clean environment are used in **assembling and inspecting** precision parts such as spindles. (Ref. LS 7) 恒温超净室用于机床主轴部件等精密部件的装配及检验。

（b）名词转译成汉语动词

If you sense an impending **collision**, press the EMERGENCY STOP button. (Ref. LS 5) 如果有可能碰撞，按下 EMERGENCY STOP（紧急停止）按钮。

（2）句子成分的转换

（a）转换成汉语主语。在英语“there be...”句型中，往往将状语略去介词而转换为汉语主语。例如：

LGMazak equips a MAZATROL system and FMS (Flexible Manufacturing System) **on every machine**, and connects them on the local network. (Ref. LS 7) LGMazak 的每一台机床都配置了 MAZATROL 系列数控系统和柔性制造系统，并将这些机床连接到公司内的局域网。

There are a variety of machine tools **in the workshop**. 这个车间有各种各样的机床。

英语中介词短语作为定语时，有时可以转换为汉语主语，同时略去介词不译。

（b）介词短语转换为定语。

The following is a list of required regular maintenance **for a horizontal machining center**. 以下是卧式加工中心的常规维护事项。

These workers first review blueprints or written specifications **for a job**. (Ref. LS 1) 首先，这些工人阅读作业零件的图纸或书面说明。

（c）转换为汉语状语从句。

There is a lot of manual intervention required to **use a drill press to drill holes**. 用台式钻床钻孔时，需要很多人工干预。

Task 1 Translate the following sentences into Chinese.

1. Check all hoses and lubrication lines for cracking.

----

2. If a bored hole is chattering because of an overextended boring bar, don't expect the machine to correct the fault.

----

3. If you sense an impending collision, press the EMERGENCY STOP button. (Ref. LS 5)

----

4. Listed are the frequency of service, capacities, and type of fluids required.

----

5. On machines with the TSC option, clean the chip basket on the coolant tank.

Task 2　Translate the following warning label on the machine into English.

机床维护警告!

1. 严禁冷却泵无水空转或反转。

2. 保持机床气源干燥（湿度小于 50%）。

3. 保持机床清洁，导轨防护罩内不允许有切屑堆积；每天应对切屑进行清除。

4. 电压要求 380V（+10%，-5%）。

5. 不得随意改动机床参数，特别是在维修说明中未提到的参数；不得随意改动机床电气连线。

6. 保证机床接地良好。

7. 禁止机床带故障工作。

8. 禁止机床使用具有腐蚀性的冷却液。

9. 禁止机床使用不符合要求的各种油液。

【引导词】run dry, run reversely, Never…/Do not…/No…/be prohibited /prevent…from, humidity, chip congestion, required, mentioned, grounding, under the circumstance of…, corrosive, unspecified

Learning Situation 7

Understand the automated factory

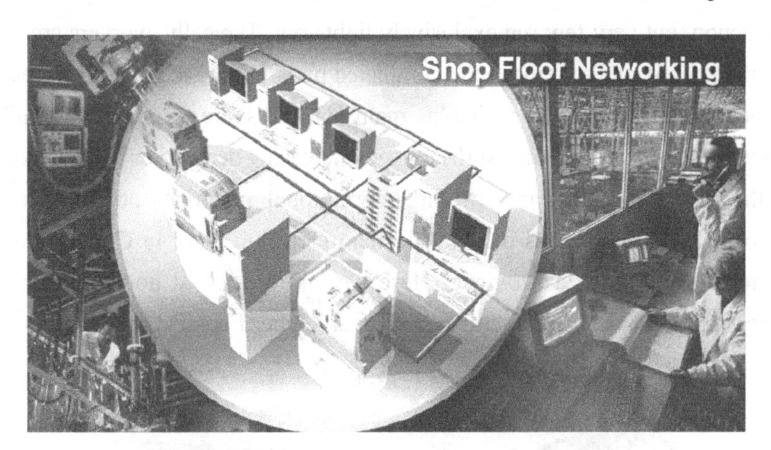

Focus of the situation

This class will show you how an automated factory operates with machine lines and management software system. [本课将展示由生产线和计算机软件系统支持的自动化工厂的运行情况。]

Field work

Search the Internet for information about the Mazak Corporation.

Part A Reading

An automatic factory is a place where raw materials enter and finished products leave with little or no human intervention. This requires CAD/CAM to program the robots, industrial robots and CNC machines to convert the raw materials to finished products, and robotic measuring machines to perform automatic inspection. An automatic factory is a computerized factory incorporating information technology and manufacturing technology.

Lights-out manufacturing means factories that run lights out are fully automated and require no human presence on-site. Thus, these factories can be run with the lights off. Many factories are capable of lights-out production, but very few run exclusively lights-out. Typically, workers are necessary to set up parts to be manufactured, and to remove the completed parts. As the technology necessary for lights-out production becomes increasingly available, many factories are beginning to utilize lights-out production between shifts (or as a separate shift) to meet increasing demand or to save money[1].

Mazak is the famous brand name of the international machine tool builder[2]. Little Giant Machine Tool Co. Ltd. (LGMazak) is the Mazak manufacturing plant in China. It was founded in May, 2000 and located in Ningxia Province. Here we take a look at how LGMazak implements factory automation and light-out manufacturing—Mazak Cyber Factory.

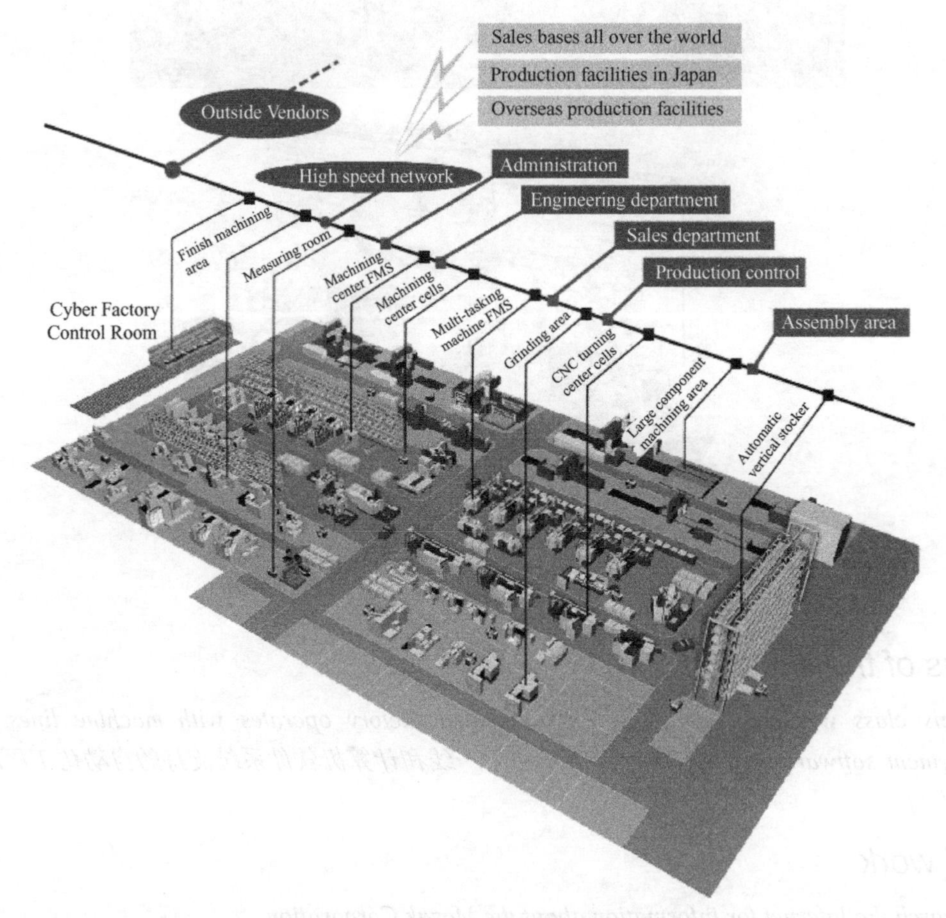

Fig. 7-1 Mazak Cyber Factory Network

High efficiency production profits from utilizing a new manufacturing concept of flexible, digital and exact production, good management realized by networking, communication, and intelligent techniques[3]. The competent management, through networking on sales, production, technique and finance, makes LGMazak the first cyber factory in China. See Fig. 7-1.

Workshop machine lines

The constant temperature workshop consists of a large parts machine line, middle and small shell parts machine line, shaft and plate parts machine line, precision machine line, unit assembly line, powder spraying line, final assembly line, automatic solid storehouse, and a precision quality test. Rooms with constant temperature and clean environment are used in assembling and inspecting of precision parts such as spindles. Advanced facilities and state-of-the-art workshop make Mazak one of the world's top ranking machine tool manufacturing corporations.

The large part machine line consisting of four sets of large scale gantry five-face machining centers is used for machining bed, column, saddle, table and other large parts. All the rough and finished machining can be done within one chuck to guarantee accurate part positioning so as to ensure the latter high precision and high efficiency assembly[4].

The shell part machine line consists of three production lines of flexible manufacturing systems (FMS). The operator edits the FMS production schedule, then the FMS production lines will operate continuously unmanned according to this schedule. High precision machines guarantee the accuracy of workpieces and eliminate human error.

The shaft and plate parts machine line consists of eight sets of the Integrex Series machines. Integrex Series multi-tasking machines totally change the traditional machining concept. The Integrex machines complete all machining processes from raw material parts to the finished parts within one chuck to guarantee machining accuracy and creates high efficiency and highly accurate machining.

The precision machine line consists of high precision horizontal machining centers and precision grinding machines. With strict temperature and humidity control, high precision key parts machining is guaranteed by high precision machines and experienced operators.

The steel plate fabrication line consists of two sets of laser cutting machines FMS, and six sets of precision NC hydraulic bending machines for processing the tool covers and coolant tanks of the machine tools.

Production support software

LGMazak equips a MAZATROL system and FMS (Flexible Manufacturing System) on every machine, and connects them on the local network. Thus all the machines can be controlled by the Cyber Production Center (CPC), the production support software, to achieve high efficiency production. See Fig. 7-2.

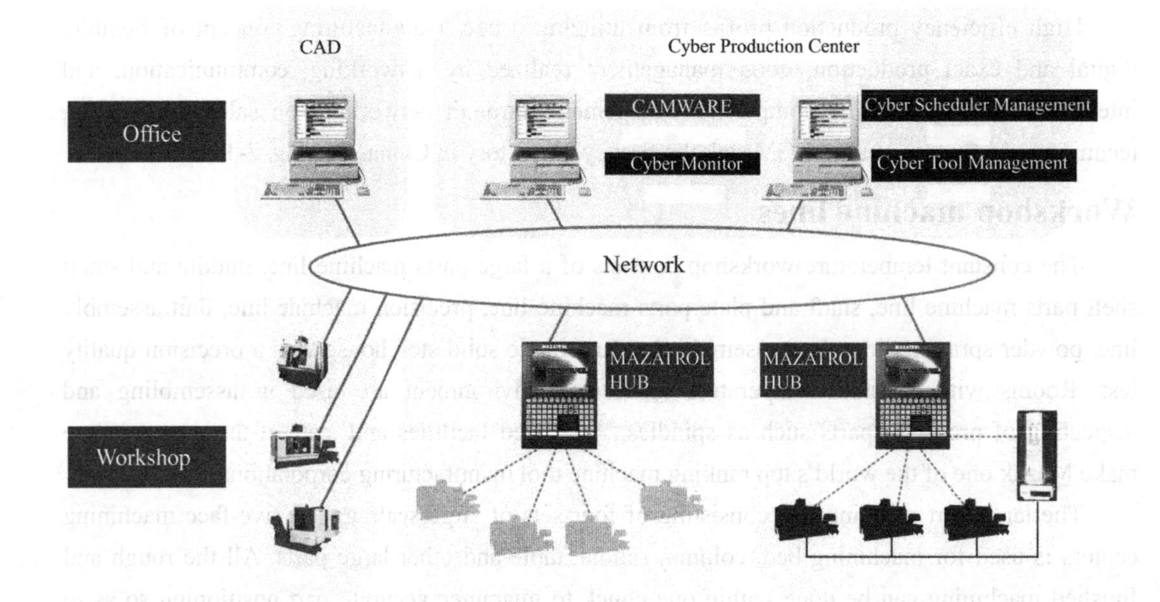

Fig. 7-2 Mazak cyber production

With Mazak CAMWARE, MAZATROL programs can be generated by a PC. CAMWARE can generate a programming file utilizing on line information for tools, fixtures, etc. to minimize the time required for programming. With the Cyber Tool Management of CPC, tool managers can monitor the state of the tools for on line machines and prepare spare tools for replacing worn tools. Production managers can prepare production schedules over Cyber Scheduler Management that are for every machine. Operators can check the production schedule and give feedback on the performance schedule of his machine over the net. Planning managers can check production conditions of the factory and make task assignment when required. Operators can check running conditions of every machine in real time over the Cyber Monitor net and work out adjustment to improve efficiency. The visible production system is formed by real time feedback and has control over the entire network. Additionally, flexible manufacturing equipment can ensure meeting the customer's delivery period requirement.

TECHNICAL WORDS

computerize	[kəm'pju:təraiz]	*v.*	用计算机处理，使计算机化
intervention	[ˌintə(:)'venʃən]	*n.*	干涉
Mazak	['meizæk, 'mæzæk]	*n.*	马扎克（公司）
cyber	['saibə]	*a.*	计算机（网络）的；信息技术的
implement	['implimənt]	*v.*	实施
cell	[sel]	*n.*	（生产）单元
fabrication	[ˌfæbri'keiʃən]	*n.*	建造；构造

PHRASES

raw material		原材料
finished product		成品
CAD/CAM		计算机辅助设计与制造
set up		装夹，调试
constant temperature workshop		恒温车间
shell part		壳体类零件
shaft part		轴类零件
plate part		盘类零件
assembly line		装配线
state-of-the-art		最先进的
large scale		大型的
gantry five face machining center		龙门式五面加工中心
Flexible Manufacturing System	(FMS)	柔性制造系统
multi-tasking machine		多任务机床，车铣复合中心
production schedule		生产计划
high precision machine		高精密机床
machining accuracy		加工精度
precision grinding machine		精密磨床
steel plate fabrication		钣金生产
laser cutting machine		激光切割机
bending machine		折弯机
tool cover		机床防护罩
local network		局域网
spare tool		备用刀具
real time		实时

NOTES

1. As the technology necessary for lights-out production becomes increasingly available, many factories are beginning to utilize lights-out production between shifts (or as a separate shift) to meet increasing demand or to save money. 随着无人值守制造技术的不断成熟，很多工厂开始在轮班之间（或单班）采用无人值守制造，以满足日益增长的需求或节省成本。

2. Mazak is the famous brand name of the international machine tool builder. 马扎克是全球知名的机床制造商品牌。*日本马扎克公司成立于1919年，主要生产CNC车床、复合车铣加工中心、立式加工中心、卧式加工中心、CNC 激光系统、FMS 柔性生产系统、CAD/CAM*

系统、CNC 装置和生产支持软件等。产品素以高速度、高精度而在行业内著称，产品遍及机械工业的各个行业。马扎克公司在全世界共有 9 家生产公司，分布于日本（5 家）、美国、英国、新加坡和中国。

3. High efficiency production profits from utilizing a new manufacturing concept of flexible, digital and exact production, good management realized by networking, communication, and intelligent techniques. 高效率的生产得益于公司采用了全新的制造概念，即加工过程柔性化、数字化、精密化，管理过程网络化、信息化、智能化。*realized by...是过去分词作为定语,说明good management。*

4. All the rough and finished machining can be done within one chuck to guarantee accurate part positioning so as to ensure the latter high precision and high efficiency assembly. 所有粗加工和精加工由一次装夹完成，保证了零件的精确定位，为后期高精密、高效率装配提供了保障。

PRACTICE

Task 1 Translate the following words into English.

1. 龙门加工中心
2. 五面加工中心
3. 卧式加工中心
4. 立式加工中心
5. 多任务机床
6. 精密磨床
7. 激光切割机
8. 数控液压折弯机

Task 2 Write the full name of the following abbreviations.

1. CAD________________________
2. CAM______________________
3. FMS________________________
4. FMC______________________
5. HMC________________________
6. VMC______________________
7. ATC________________________
8. APC______________________

Task 3 Fill in the brackets with words that have similar meaning to the underlined words, changing their forms if necessary.

1. (   ) The competent management, through networking on sales, production, technique and finance, makes Mazak a <u>cyber</u> factory.

2. (   ) High efficiency production profits from <u>utilizing</u> a new manufacturing concept.

3. (   ) Inspecting precision parts such spindles, advanced facilities and <u>state-of-the-art</u> workshop make Mazak one of world's top ranking machine tool manufacturing corporations.

4. (   ) All the rough and finished machining can be done within one <u>chuck</u> to guarantee accurate part positioning.

5. (   ) The Integrex machines complete all machining processes from raw

material parts to the finished parts within one chuck to <u>guarantee</u> machining accuracy and creates high efficiency and highly accurate machining.

6. (　　　　　) Flexible manufacturing equipment can ensure meeting the customer delivery period <u>requirement</u>.

7. (　　　　　) High efficiency production profits from utilizing a new manufacturing concept, flexible, digital and <u>exact</u> production.

8. (　　　　　) High precison machines guarantee the accuracy of workpeices and <u>eliminate</u> human error.

9. (　　　　　) Planning managers can check production conditions of the factory and make task assignment <u>when required</u>.

10. (　　　　　) Mazak is the famous brand name of the international machine tool <u>builder</u>.

Task 4 Match the transitive verbs with the objects in the column on the right.

Transitive verbs	Objects
guarantee	program
meet	concept
perform	accuracy
generate	machine
utilize	requirement
manufacture	inspection
control	part

Task 5 List the characteristics of lights-out manufacturing in the following blanks. You can also write down your own opinions.

1. Employ automated parts loaders

2.

3.

4.

5.

6.

Part B Listening

Task 1 Listen to the five statements twice and write them down.

1. --

2. --

3. ____

4. ____

5. ____

Task 2 The following video is about Mazak Cyber Production Center. Watch it first, then listen to it twice and fill in the blanks with what you hear.

LGMazak equips a (1)________ system and an FMS ((2)________________________) on every machine, and connects them on the (3)________________. Thus all the machines can be controlled by the Cyber Production Center (CPC) to achieve high efficiency production.

With Mazak CAMWARE, MAZATROL programs can be generated by a (4)________________. (5)________ can generate a programming file utilizing on line information for tools, fixtures, etc. to minimize the time required for programming. With the (6)________________ of CPC, tool managers can monitor the state of the tools for on line machines and prepare spare tools for replacing (7)________________ tools. Production managers can prepare production schedules over (8)________________________ that are for every machine. Operators can check the production schedule and give feedback on the performance schedule of his machine over the net. Planning managers can check production conditions of the factory and make task assignment when required. Operators can check running conditions of very machine in real time over the (9)____________ net and work out adjustment to improve efficiency. The visible production system is formed by (10)____________ feedback and has control over the entire network. Additionally, flexible manufacturing equipment can ensure meeting the customer's delivery period requirement.

Task 3 Watch and listen to the above video once more, and choose the best answer to each of the following questions.

1. How many job titles are mentioned in the video?

(A) Three. (B) Four. (C) Five.

Your answer: ____________

2. Who prepare production schedules?

(A) Planning managers.

(B) Programmers.

(C) Production managers.

Your answer: ____________

3. Who check production conditions of the factory and make task assignment?

(A) Planning managers.

(B) Programmers.

(C) Production managers.

Your answer: ____________

4. Who check running conditions of every machine?

(A) Tool managers.

(B) Operators.

(C) Production managers.

Your answer: ___________

5. Mazak Cyber Production Center is:

(A) A large workshop.

(B) A multi-tasking machining center.

(C) A computer software system.

Your answer: ___________

Part C Writing

如何写英文摘要

毕业论文需要撰写英文摘要（Abstract）和关键词（Key Words），实际工作中发表的科技论文都要附英文标题和摘要，以便于国际学术交流和检索。摘要本质上就是一篇高度浓缩的论文，一般不超过300个词，由于大多数检索系统只收录论文的摘要部分或其数据库中只有摘要部分免费提供，并且有些读者只阅读摘要而不读全文或经常根据摘要来判断是否需要阅读全文，因此，摘要的清楚表达十分重要。好的英文摘要对于增加期刊和论文的被检索和引用机会起着不可忽视的作用。摘要的翻译要从总体结构上把握大意，语言的组织应简明扼要，有实际内容，合乎英语语法。

摘要在内容上大致包括背景（Background）、方法（Methods）、结果（Results）、结论（Conclusions）四个方面。

1. 摘要写作的时态

如果句子的内容是不受时间影响的普遍事实，应使用现在时态；如果句子的内容是对某种研究趋势的概述，则使用现在完成时态。概述实验程序、方法和主要结果，通常用现在时态；用于叙述过去某一时刻(时段)的发现、某一研究过程(实验、观察、调查等过程)，则用过去时态。

2. 摘要写作的语态和人称

由于主动语态的表达更为准确，且更易阅读，因而目前大多数期刊都提倡使用主动语态，第一人称和第三人称均可。

3. 摘要的常用表达方法

由于摘要的英文表达要求用词简明、层次清楚，因此，掌握一些特定的规范表达对于摘要的撰写是很有帮助的。

（1）回顾研究背景。常用词汇有：review, summarize, present, outline, describe 等。

（2）阐明写作或研究目的。常用词汇有：purpose, attempt, aim, intend 等。另外还可以用动词不定式充当目的状语来表达。

（3）介绍论文的重点内容或研究范围。常用的词汇有：study, present, include, focus,

emphasize, emphasis, attention 等。

（4）介绍研究或试验过程。常用词汇有：test, study, investigate, examine, experiment, discuss, consider, analyze, analysis 等。

（5）介绍应用、用途。常用词汇有：use, apply, application 等。

（6）展示研究结果。常用词汇有：show, result, present 等。

（7）介绍结论。常用词汇有：summary, introduce, conclude 等。

4. 摘要中的缩略语、简称、代号

除了相邻专业的读者也能清楚理解的以外，在首次出现时必须加以说明。

5. 摘要翻译切忌逐词逐句对应

大部分论文作者不能做到用英语思考（think in English），往往只是根据中文进行翻译。事实上，由于两种语言表达方式的差异以及中英文摘要读者对象的不同，中英文摘要不必强求逐词逐句一致，而应从总体上把握摘要的大意。

6. 充分利用网络搜索对应的英文词汇

当遇到专业名词不会表达时，只要在搜索栏内输入该专业名词的中文+有关的英文线索即可。例如，要查“CA6140 车床拨叉零件的机械加工工艺及夹具设计”中“拨叉零件”的专业表达法，但只知道零件是“parts”，因此输入“拨叉零件 parts”，即可找到答案。

摘要样例：

配置 FANUC 系统的数控机床的调试

Commissioning of CNC machine with FANUC control

摘　要：随着数控机床在我国制造业的普及使用，研究其安装调试过程及故障维修方法显得日趋重要。本文以配置品牌 FANUC 0*i* 系统的加工中心的调试为例，介绍了配以 FANUC 系统的数控机床的一般调试步骤，同时也说明了数控加工中心各功能动作的实现过程。本文对机床大修或旧机床数控化改造具有一定的指导意义。

Abstract: With the wide use of CNC machine tool in Chinese manufacturing industry, more importance has been attached to its commissioning and troubleshooting.（以上为背景介绍）This article presents commissioning of machining center equipped with widely used brand FANUC 0*i* control and describes how machine functions are implemented. These descriptions show the general commissioning procedure of the machine tool with FANUC control.（以上为论文的重点内容） We intend to give directions to machine tool overhaul and retrofitting. （以上为论文的目的）

关键词：数控机床　FANUC 0*i*系统　加工中心　调试过程　数控化改造

Key words: CNC machine tool, FANUC 0*i* control, machining center, commissioning procedure, CNC retrofitting

Task　Translate the following abstract into English.

细长轴的车削加工

摘要：在普通车床车削加工零件中，细长轴类零件的加工是个难题，这是因为细长轴的刚性差，加工时工件极易弯曲和振动，没有合理的措施和办法是难以加工的。本文依据金属切削的原理，结合细长轴类加工件的特性，在总结细长轴类零件车削加工技术应用的生产实践经验的基础上，提出了细长轴类零件车削加工的解决方案，使之符合车削加工的工艺技术要求，提高了加工件的成品率，进而降低了生产成本，提高了生产效益。

关键词：细长轴　车削加工　普通车床

Abstract: ____

____

____

____

____

____

____

____

Key words: ____

____

Part D　Grammar and Translation

科技英语翻译小技巧

译文应忠实于原文，准确地、完整地表达原文的内容。译者不得随意对原文的思想加以歪曲、删除，也不得有遗漏和篡改。译文语言必须符合规范，符合译文民族语言的习惯；要用民族的、科学的、大众的语言来表达原文的思想内容，以求通顺、畅达。对于技术资料的翻译，应努力使译文做到“准确、通顺、易懂”，尤其是准确，它是科技翻译的灵魂。这就要求译者应具备相关行业的专业知识，熟悉行业术语，运用逻辑判断，才能成就准确的译文。

1．翻译步骤

拿到一篇内容熟悉或不熟悉的英文，翻译时不可能一气呵成，必须斟字酌句，反复阅读，反复修改。步骤如下：

（1）快速通读全文（Scanning），获得大意；

（2）确定关键词在本专业中的含义；

（3）副词、介词、定语成分先不急着翻译，写下各分句主干成分的中文意思（打草稿）；

（4）根据副词、介词、状语、定语成分的含义，添加文章中的补充说明部分；

（5）读几遍，看是否通顺、易于理解，并符合汉语表达习惯；

（6）修改（意译），定稿。

2．几个注意点

（1）翻译专业英语时，要结合自己的专业知识（选词），提炼其中的含义，翻译成既符合汉语的习惯，同时又符合本行业术语习惯的文字。

Because **the front edge of the tool** is ground at **an angle**, the left side of the tip should engage the work, but not the entire front edge of the tool. (Ref. LS 3) 由于刀具的前部刀刃（**副切削刃**）被磨出一个角度（**副偏角**），刀尖的左侧应该切入到工件，而不是整个前部刀刃（副切削刃）与工件接触。

（2）对并列内容的翻译，注意中文的排比，如 LS 2 中：

In general, a lathe apron contains the following mechanical parts:

(a) A carriage handwheel for moving the carriage by hand along the bed. …

(b) Gear trains driven by the feed rod.

(c) A selective feed lever, here called the longitudinal& cross feed lever, is provided for engaging the longitudinal feed or cross feed as desired.

(d) Friction clutches operated by the longitudinal& cross feed lever on the apron are used to engage or disengage the power feed mechanism.

(e) A selective lever for opening or closing the half-nuts, called half-nut lever.

(f) Half-nuts for engaging and disengaging the lead screw when the lathe is used to cut threads.

译文如下：

总之，车床溜板箱包含了下列机械部件：

(a) 手动沿床身移动大拖板的大拖板手轮……

(b) 由光杠驱动的齿轮系……

(c) 进给选择手柄（这里称为纵向/横向进给手柄），用于按需进行纵向进给或者横向进给的齿轮啮合。

(d) 由溜板箱上的纵向/横向进给手柄操作的摩擦离合器，用于啮合或断开自动进给机构。有些车床具有独立的离合器分别用于纵向进给和横向进给，有些则只使用一个离合器，用于两个方向的进给。

(e) 开合半开螺母的选择手柄，称之为开合螺母手柄。

(f) 车床螺纹切削时用于啮合和断开丝杠的开合螺母……

（3）几个常用词的译法：

① 对于由动词加 er 或 or 构成的名词，常可译成“……器，……装置，……机”。如：autoloader（自动上下料装置）, regulator（调节器），conveyor（传送装置）。

② unit 的译法：可译成“装置，组件，单元，环节”。

③ condition 的译法：可译成“工况”。

④ be designed to 往往无须翻译出来。如：

Many forms of CNC machines are designed to enhance or replace what is currently being done with more conventional machines. (Ref. LS 4) 很多数控机床可以增强或替代普通机床的功能。

⑤ allow 的译法：可译成“允许，使……可以……，用于……，具有……的功

能”。如：

EDIT mode allows manual editing changes in a program or creation of a new program. (Ref. LS 5) EDIT（编辑）方式用于手动修改程序或建立新程序。

（4）掌握常用专业词汇的中文表达法。如：compound（小拖板），ball screw（滚珠丝杠），carbide tool（硬质合金刀具）。

（5）在理解英文意思的基础上，在不背离英文含义的原则下，怎么通顺怎么翻译。

Task　Read the following text and translate it into Chinese.

1. Integrex Series multi-tasking machines represent the full integration of machining centers and turning centers, allowing diverse families of parts（零件族） to be machined in one setup.

----

2. CNC stands for computer numerical control. It allows you to control the motion of tools and parts through computer programs that use numeric data. CNC can be used with nearly any traditional machine. The most common CNC machines found in the machine shop include machining centers (mills) and turning centers (lathes).

Part E　Supplementary Reading

Mazak's Integrex/Palletech Cell

Integrex Series multi-tasking machines represent the full integration of machining centers and turning centers, allowing diverse families of parts（零件族） to be machined in one setup. It's what we call done-in-one. Mazak focused Integrex technology on a portion of our machining operations that manufactures over 80 distinct part numbers, including spindle shafts and chucks for machine tools. Four Integrex machines with 120-tool capacity, 12" chucks, and automatic chuck jaw changers are integrated with a 31-pallet cell that moves workpieces between machines, load/unload stations, and queue stations. Moreover, this 4-machine cell replaces eleven machine tools including three HMCs, two VMCs, two turning centers, four multi-tasking machines, and all of the tooling that went with them.

Palletech palletized（用托盘装运） machining systems are pre-engineered automation that can serve one machining center with a few pallets – or up to eight machines with one hundred pallets. And, customers can add machines, pallets and load/unload stations as needed. Palletech includes a simple, yet effective scheduler, which allows jobs to be queued and prioritized（优先处理）for lights-out operation. Yet when the prototype job or the emergency delivery arises, as it always does, it can be given top priority in the schedule without the need for re-organizing other work.

Palletech cells are an effective strategy for both low-volume and high-volume requirements. These cells are compatible with horizontal machining centers, 5-axis machining centers, plus the ultimate multi-tasking machines, the Integrex Series.

Fig.7-3 shows a variety of Palletech configurations.

Two-level stocker for more efficient pallet storage

Lights-out production of batch parts or individual components

Load/unload stations with 90° indexing

Multi-tasking machines

Fig. 7-3 Variety of Palletech configurations

Learning Situation 8

At CIMT

Focus of the situation

CIMT is one of the four most important machine tool exhibitions in the world. Engineers and managers from all parts of the world flock to Beijing, where they get abreast of the latest manufacturing technology and promote their new products. Also, some people may find job opportunities. [中国国际机床展览会是世界四大机床展览会之一。世界各国的工程师和经理云集北京，了解世界制造技术新动态，推销最先进的机床制造产品。同时，有人也会发现一些工作机会。]

Field work

Visit http://www.imts.com/ for information about the international machine tool show.

Part A Reading

Outside the hall...

Li: Welcome to the China International Machine Tool Show. The opening ceremonies have just begun.

Brown: Yes, the scene is lively.

Li: Look at those large balloons in the air with welcoming slogans on them.

Brown: They are very impressive indeed. There seems to be a big turnout.

Li: Exactly. The registration is around 45,000 visitors so far, from all parts of the world and the number is expected to increase over the next two or three days.

Brown: I'm lucky to have this opportunity.

Li: You are right. The show has been very important to China's machine tool industry. There are about 1,000 new machine tools on display. Delegations representing many different countries or regions are participating in this CIMT.

Brown: Fantastic! Do you have such a big event every year?

Li: Actually, it has been held in China every 2 years ever since 1989. And the show has been recognized as one of the top four marketing activities in the world's machine tool sector.

Brown: I'm sure to come next time.

Li: You are welcome!

Inside the hall...

Brown: Look! Aren't the exhibits spectacular?

Li: Sure. This is the exhibition hall 8A. It's divided into four sections. Here on display are some new domestic machine tools. Many of them have caught up with the technical levels of similar products made abroad. Let me show you around.

Brown: It's very kind of you! Oh, that's a big simultaneous five-axis CNC machine.

Li: Yes. This new machine reaches the advanced world level. It is suitable for the aviation industry.

Brown: I see.

Brown: Is this a boring-milling machine?

Li: That's right. It is highly recommended in the world because it is economical, easy to operate and outstanding in performance.

Brown: That sounds interesting. But the size is a bit small.

Li: It is specially designed to handle small pieces, and it is stable and efficient. The bigger one is also available. Over there, you see.

Brown: What's the unit price?

Li: Here is our price list and this is the catalogue.

In the negotiation booth...

Brown: We have studied your catalogue and we have great interest in your boring-milling machine.

But by our calculations your price appears a little high.

Li: What quantity are you looking to order?

Brown: We plan to order 15 units, provided the price is right.

Li: To be frank, this machine was out of stock for a while because the demand exceeded the supply. The list price remains unchanged. But because of the large order size, and because we would like to establish a long-term relationship with you, we are prepared to offer you a reduced unit price of $20,000 FOB Shanghai. That's our bottom price.

Brown: It seems acceptable. What is your lead time?

Li: Twelve weeks.The goods can be ready by August.

Brown: Well, we expect to use them this October. Time is too tight. We need to transit the goods at Singapore since there is no direct vessel from Shanghai to Lagos. Could you get the goods ready for shipment by mid July?

Li: July is OK.

Brown: When can we sign the contract?

Li: Tomorrow afternoon.

Brown: See you tomorrow then.

Li: See you tomorrow.

TECHNICAL WORDS

slogan	['sləugən]	*n.*	标语，口号
turnout	['tə:n,aut]	*n.*	出席者，一项活动的到场人数
opportunity	[ɔpə'tju:niti]	*n.*	机会
delegation	[deli'geiʃ(ə)n]	*n.*	代表团，展团
exhibit	[ig'zibit]	*n.*	展品
spectacular	[spek'tækjulə(r)]	*a.*	壮观的
negotiation	[ni,gəuʃi'eiʃ(ə)n]	*n.*	谈判
booth	[bu:θ]	*n.*	展位，摊位
catalogue	['kætəlɔg]	*n.*	产品样本，目录
order	['ɔ:də(r)]	*v. &n.*	订货，订购
vessel	['vesl]	*n.*	船，舰
shipment	['ʃipmənt]	*n.*	装船，交货
contract	['kɔntrækt]	*n.*	合同
transit	['trænsit, 'trɑ:-]	*v.*	运输

PHRASES

opening ceremony	开幕式
participate in	参加

exhibition hall	展馆，展厅
domestic machine tool	国产机床
simultaneous five-axis CNC machine	五轴联动数控机床
boring-milling machine	镗铣床
unit price	单价
out of stock	脱销
lead time	交货时间
Lagos	拉各斯，尼日利亚首都
direct vessel	直达航运
sign the contract	签定合同

NOTES

1. CIMT 是 China International Machine Tool 的缩写，代表中国国际机床展览会，是世界四大国际机床名展之一，其他国际机床展览会分别在美国（称为 IMTS）、欧洲（称为 EMO）和日本（称为 JIMTOF）举行。首届中国国际机床展览会于 1989 年在上海举行，以后每两年一届。第十一届中国国际机床展览会 CIMT'2009 于 2009 年 4 月 6 日至 11 日在北京举行。

2. FOB 是 Free on Board 的缩写，离岸价。CIF 是 Cost, Insurance &Freight 的缩写，成本、保险加运费价。这些都是国际贸易术语。

Part B Listening

Task 1 The following video is an IMTS'2010 showroom. Here the exhibitor is Absolute Machine Tools, Inc., an American machine tool importer and distributor. Watch it first, then listen to it twice and complete the following tables according to what you hear and see. Then translate your answers into Chinese orally. Some have been given.

Table 8-1

No.	Variety of industries related	Service provided
1	Aerospace	Installation
2		
3		
4		

Table 8-2

Brand name	Type of machine tool
Tongtai	HG-1250 Large Capacity HMC
YOU JI	
JOHNFORD	
Quick-Tech	

Task 2 Watch the above video again, and then decide whether the following statements are true (T) or false (F).

1. (   ) Absolute Machine Tools is an American MTB.
2. (   ) IMTS'2010 was held in Chicago, the U.S, in 2010.
3. (   ) Both Tongtai and YOU JI provide vertical turning machines.
4. (   ) If you need purchase a vertical machining center, you may choose YOU JI or JOHNFORD.
5. (   ) Only JOHNFORD provides double column machining centers.

Task 3 Write out the full form for each of the following abbreviations according to the above video. Fig.8-1 shows the logo of IMTS'2010.

No.	Abbreviation	Full form
1	IMTS	
2	HMC	
3	VMC	
4	DMC	
5	VTC	
6	VTL	
7	ATC	
8	APC	

Fig.8-1 Logo of IMTS'2010

Task 4 The following video gives you tips during a job interview. Watch it first, then listen to it twice and complete the following table according to what you hear and see.

No.	Interview tips of *should*	Interview tips of *shouldn't*
1	Know what days and hours you are available to work	Talk on the phone or text
2		
3		
4		
5		
6		
7		
8		
9		
10		

【语音提示】

Vocabulary		Vocabulary	
interview(er)	面试（者）	perfume	香水
resume	简历	text	发短信

续表

Vocabulary		Vocabulary	
reference	证明人	chew gum	嚼口香糖
portfolio	文件夹	slouch	无精打采地坐（站）
distract	分散（注意力）	slang	俚语
notepad	便条簿	swore	脏话

Part C Speaking

Task 1 ***Work in pairs. You work for a company as a salesman. And you are in your booth at CIMT. Some foreign businessmen come up to make some inquiries about your product and the price. Make a conversation between the foreigner(s) and you according to the situation. Role-play the conversation.***

Task 2 ***Work in pairs. Create a scene of interviewing with a foreign company. Role-play the interview.***

WORKING SITUATION

You will begin job hunting in the near future. To be a successful job seeker, you need to prepare for the **job interview with the foreign company**.

【引导文】面试常见问题及答案建议 Common interview questions and answers

1. Tell me about yourself.

The most often asked question in interviews. You need to have a short statement prepared in your mind. Be careful that it does not sound rehearsed. Limit it to work-related items unless instructed otherwise. Talk about things you have done and jobs you have held that relate to the position you are interviewing for. Start with the item farthest back and work up to the present.

2. What experience do you have in this field?

Speak about specifics that relate to the position you are applying for. If you do not have specific experience, get as close as you can.

3. Do you consider yourself successful?

You should always answer yes and briefly explain why. A good explanation is that you have set goals, and you have met some and are on track to achieve the others.

4. What do classmates say about you?

Be prepared with a quote or two from classmates. Either a specific statement or a paraphrase（解释）will work.

5. What have you done to improve your knowledge in the last year?

Try to include improvement activities that relate to the job. A wide variety of activities can be mentioned as positive self-improvement. Have some good ones handy to mention.

6. Why do you want to work for this organization?

This may take some thought and certainly, should be based on the research you have done on the organization. Sincerity is extremely important here and will easily be sensed. Relate it to your long-term career goals.

7. What kind of salary do you need?

A loaded question. A nasty little game that you will probably lose if you answer first. So, do not answer it. Instead, say something like, That's a tough question. Can you tell me the range for this position? In most cases, the interviewer, taken off guard, will tell you. If not, say that it can depend on the details of the job. Then give a wide range.

8. How long would you expect to work for us if hired?

Specifics here are not good. Something like this should work: I'd like it to be a long time. Or as long as we both feel I'm doing a good job.

9. Why do you think you would do well at this job?

Give several reasons and include skills, experience and interest.

10. Are you a team player?

You are, of course, a team player. Be sure to have examples ready. Specifics that show you often perform for the good of the team rather than for yourself are good evidence of your team attitude. Do not brag（自夸）, just say it in a matter-of-fact tone. This is a key point.

Notes

Part D Writing

如何写英文求职书

1. Resume（个人简历）

A resume is a summary of your personal background and your qualification for a job or enrolling at a school. It helps your potential employer or supervisor see at a glance whether you are suited for the job opening or qualified for a certain position. The items in a resume usually include Name, Address, Telephone Number, Date of Birth, Marital Status, Citizenship, Education, Work Experience, Languages, Interests, References, etc. Generally you need a bilingual resume.

Sample resume:

Name	Zhou Han	Sex	Male	Date of Birth	July 1989
Height	1.73m	Native Place	Changzhou, Jiangsu	Hobbies	Traveling, Photograph, Swimming
Weight	65kg	Health	Excellent	Mobile Phone	15106113993
Objective	Machinist				
Education	Sept. 2009–July 2012 Changzhou Institute of Mechatronic Technology, Diploma of technical college				
Major	Mechanical Manufacturing and Automation				
Practice	Making parts by hand, making parts by conventional turning, making parts by CNC machining, designing gear decelerators, designing fixtures, measuring and plotting milling cutters, AutoCAD training				
Work Experience	July 2010–Sept. 2011 Changzhou Sanli Precision Machine Co. Ltd., Operator of CNC lathe				
Rewards	Sept. 2009 Bosch Rexroth Scholarship Dec. 2009 First-class Prize in National English Contest for College Students Mar. 2010 First-class Scholarship				
Certificates	Mar. 2009 Certificate in CET-Band 4 May 2009 Certificate in Middle Level of 2D CAD July 2009 Certificate in Middle Level of Lathe Machinist May 2010 Certificate in Middle Level of CNC Lathe Machinist				
Courses	Mechanical drafting, Mechanical manufacturing basics, Mechanical manufacturing process and apparatus, Electrical control and PLC, Hydraulic and pneumatic drive technology, NC programming				
Life Motto	No pains, no gains.				

2. Application letter（申请信）

In an application letter, you should usually include the following:

- Job or position wanted, or purpose of writing (brief, in opening paragraph)
- Qualifications (briefing education and experience)
- Achievements
- Skills
- Interests and outside class activities that give a general picture about yourself
- Anything that might help the reader decide if you are the kind of person they want

Sample application letter:

Dear Manager:

I would like to be considered as a candidate for the position you are advertising. I am currently finishing my study in Mechanical Manufacturing and Automation at Changzhou Institue of Mechatronic Technology. I have taken every mechanical course offered at the college and have a solid background in the following skills: part handmaking, engine lathe turning and CNC machining.

My knowledge of machining goes beyond my formal classroom education. For the past two years I have worked part-time at Sanli Precision Machine Co., where I have gained experience in process planning and precision control. Also, on my own initiative, I participated in designing a gear decelerator for the company.

I have always had the dream of becoming an employee of a foreign company. So I have applied myself to learning English well. Now I have passed CET-band 4.

I value team work spirit and possess amiable and enthusiastic personality. I have fast-learning, analytical and creative abilities.

In short, I believe I have the up-to-date mechanical background and professional drive needed to contribute to your company. I have enclosed a copy of my resume to give you further details about my experience. I look forward to speaking with you then.

Sincerely yours,

Zhou Han

Task Write your own bilingual resume and application letter for your targeted job position. Or you may apply for the positions given below offered by a foreign manufacturing company.

Position 1: Equipment Mechanic

Requirements:

1. Technical educational background (major in machining, maintenance or similar major)
2. Knowledge of pneumatic and hydraulic systems regarding the electrical side
3. Preferably product knowledge of generator gearboxes
4. Ability to work independently
5. Flexibility to be sent to Germany for a training period of several months
6. Flexibility concerning shift work
7. English skills spoken/written appreciated

Position 2: Equipment Electrician

Requirements:

1. Technical educational background (major in machinery & electronics, maintenance or similar major)

2. 1-2 years or more work experience in maintenance of CNC machines, assembly and repair work

3. Knowledge of pneumatic and hydraulic systems regarding the electrical side

4. Preferably product knowledge of generator gearboxes

5. Ability to work independently

6. Flexibility to be sent to Germany for a training period of several months

7. Flexibility concerning shift work

8. English skills spoken/written appreciated

Resume

Application letter

Part E Supplementary Reading

At a dinner party

Mr. Hall (H)

Mr. Chen (C)

Mr. Deng (D), director of Shanghai Bureau of Machine Building Industry

Mr. Bob Lee(L) and Mr. Charles Smith(S), both engineers

H: Hello Mr. Deng, my old friend. Nice to see you again. How are you?

D: Very well, thank you, Mr. Hall. What a nice surprise to meet you again so soon.

H: Do you remember that at the farewell meeting in San Francisco I promised to come to China to discuss a joint venture（合资）with the No. 2 Machine Tool Plant（第二机床厂）?

D: Yes, you said that. I understand that things are going on successfully.

H: Yes, very much so, Mr. Deng. Your people, especially Mr. Chen, have given us every assistance and made things very easy for us.

D: It has been nearly two weeks since our American friends came to our city. It is worth noting that through our mutual cooperation, we have succeeded during such a short span of time, in concluding an agreement about the joint venture project, and as the discussions are still on going, we expect that a number of long-term arrangements will be made and a number of general questions which have a close bearing on the project will be discussed this time. We are convinced that through these personal contacts both sides have acquired a better knowledge of our mutual requirements. The conclusion of the agreement, we hope, will mark the beginning of our mutual endeavors, placing our mutual business on a more durable footing.

H: For the success of our preliminary talks（初步会谈）, we are indebted （感恩） to director Deng and Mr. Chen. We thank you very much. In our discussion on general issues about the project, we have met with a forthrightness（坦率）, an understanding, and a cooperation that has already eliminated many difficulties and promises to solve the remaining problems soon.

D: Now, shall we go to the table?

C: Mr. Hall, take this seat, please, and Mr. Lee, here, please.

H: Thank you.

D: I am happy to host this dinner party in honor of our American friends who have done so much to promote our cooperation.

H: We are very honored to be invited to this magnificent dinner.

C: Mr. Hall, what would you like to drink, beer or orange juice?

H: Beer, please.

C: What about you, Mr. Lee?

L: I'll have orange juice.

S: What's in this small glass?

D: It's Mao-tai, a real Chinese specialty. You must try some later.

S: I have heard so much about Mao-tai. I'm afraid it might be too strong for me.

D: Yes, it's a bit strong, but it won't go to your head. Mr. Hall has been to China three times. He knows that.

D: When it comes to a good drink, nothing can beat Mao-tai. Some visitors say that it's a shame to leave China without doing three things - going to the Great Wall, eating Peking duck and drinking Mao-tai.

L: If that is the case, I'll take a sip later.

D: Now, shall we start the ball rolling（开头动筷）? Let's have the cold dish first.

S: A cold dish? But it looks like a peacock to me, I thought it was a decoration.

D: It's an assorted cold dish arranged in the shape of a peacock. Help yourselves to what you like on the plate - ham, chicken, duck, beef, egg, and mushroom.

L: It's so beautiful that I hate to destroy it.

S: Hold it. Let me take a picture first.

L: Chinese cooks are great artists, aren't they?

D: The eight small plates around the big one are hors d 'oeuvres（冷盘）. Just revolve the center of the table to the desired position.

D: I notice that you are quite at home with chopsticks, Mr. Hall.

H: Back in my country, I often go to Chinese restaurants and eat with chopsticks. I have gradually mastered the skill.

L: "When in Rome do as the Romans do. " I also like to use chopsticks although I can't get along with them very well as I only began to handle them after I came to China on this trip.

D: Since you are old customers in Chinese restaurants at home, you probably know the characteristics of Chinese cooking.

H: No, I don't, but what I do know is that Chinese dishes are delicious.

D: You see, real Chinese cuisine places stress on color, smell, taste, and sound.

L: The dishes we have had so far certainly look beautiful, smell good, and taste nice. But what about the sound?

D: You will know in a minute.

S: What's the waiter bringing in this time?

C: It's our next course called "Shrimp with Fried Rice Crust（虾仁锅巴）".

L: Now, he is pouring the shrimps over the rice crust.

C: Hear the sound?

L: Now, I understand why you include sound in the features of Chinese cooking.

H: Mm. The shrimps are great.

C: Here comes the four-colored vegetable with Chinese cabbage, mushrooms, fungus, and bamboo shoots.

L: What beautiful colors. They look inviting.

S: Chinese are great vegetarians, we believe.

D: Yes, vegetables are our main diet. Shanghai is self-sufficient in vegetables.

L: We have learned that you city people buy vegetables in the market every day.

D: Yes, we like to eat our vegetables fresh.

L: I think this is what makes you people so healthy.

S: You see, we Americans get old early because we eat too much fat and canned food but not enough vegetables.

H: And we don't do as much exercise as you.

S: These days I get up very early and take a walk along the waterfront. I see many people doing exercises. They all look very healthy.

C: As your saying puts it, " A sound mind in a sound body（心宽体胖）. "

D: Friends, you mustn't forget your drink. Here's to your health.

H, L, and S: To your health.

D: Gan Bei.

L: What does that mean?

H: It is the equivalent of our "bottoms up. "

C: You have other similar expressions like "Mud in your eye, " " Down the hatch, " " Skin off your nose. " (*Laughter*)

L: What's the stuff wrapped in leaves?

C: It's Duck Wrapped in Lotus Leaves（荷叶包鸭）.

H: It tastes different from the duck I had in Beijing last time.

D: It was cooked in a different way. In China, the cuisine varies from place to place. But now there is a tendency for mutual adaptations among the different styles. In Beijing, sometimes, the whole banquet has dishes taken from different parts of a duck.

H: I was told that you can eat the whole duck except the quack.

D: That's true. Now, Mr. Lee, you have stopped eating.

L: I am already up to here. (*Pointing his throat*)

K: You haven't eaten much either, Mr. Smith. If you go hungry, it's your responsibility. (*Laughter*)

S: The menu tells me that we are only half way through. I am saving room for the rest.

D: Mr. Hall. Let me refill your glass, the white one.

H: No more for me, thanks.

D: Why, I know you can hold your liquor.

H: I have already acted upon the principle of "from each according to his ability" tonight. Two more glasses of Mao-tai will surely make me drunk as a lord.

D: Let's drink the table wine then. To our friendship.

H, L, and S: To friendship.

H: Speaking of friendship, I feel that the Chinese are the friendliest people I've ever met. Everywhere we go they greet us with smiling faces and are always ready to help.

D: Thank you for speaking so highly of our people and of what we have done. But the other side of the picture is that there are still a lot of problems to be solved and we have many shortcomings. We have to make greater effort to get rid of the backwardness of our country and catch up with and surpass advanced world levels.

H: "Rome was not built in a day. " Your country has already made great strides. You will surely achieve the four modernizations.

D: We Chinese people are marching confidently towards that grand goal under the leadership of the Communist Party Central Committee. To realize the four modernizations, we not only should rely on the efforts of our own people, but also need help from the people of other countries. Such help includes criticisms and suggestions, and so forth.

H: It seems Mr. Deng is asking us for our criticisms and suggestions. Do you have any?

L: I have one complaint: you feed us too well, we have all put on weight.

D: My criticism is that you have taken too good care of us. You have spoiled us.

D: You are our special guests. You should be treated that way. If you go home lean, your families will settle the score（算账）with us and nobody will dare to come to China again. (*Laughter*)

L: Look, the waiter is coming with another dish.

D: That's our soup served in a "bowl" made of a winter melon.

S: The soup is super.

L: Will there be still more courses after the soup?

D: Not really. In a Chinese dinner, soup is served towards the end of a meal while you start your meal with it.

S: So I would say that I enjoy the dinner from nuts to soup, instead of "from soup to nuts." We thank you, Mr. Deng, and Mr. Chen for this wonderful meal.

D: Finally, I would like to propose a toast to the friendship between us and the health of you all. Bottoms up.

L, S, and C: Bottoms up.

Assessment 1

Ⅰ. Decide whether the following statements are true(T) or false(F). *(10×1%)*

1. ( ) Non-cylindrical shapes can be faced on the lathe.

2. ( ) The finishing cut can be made at lower RPM.

3. ( ) To obtain uniform cutting speed, the lathe spindle must be revolved slower for workpieces of small diameters and faster for workpieces of large diameters.

4. ( ) Generally, the deeper the cut, the slower the speed, since a deep cut requires more power.

5. ( ) The NC program can be run only once.

6. ( ) Turning with power feed will produce more even finish than is generally achievable by hand feed.

7. ( ) The work zero point of the part is decided by the programmer.

8. ( ) The X-axis moves the table to and from the operator on the vertical machining center.

9. ( ) Micrometers are more accurate than calipers.

10. ( ) Drilling and tapping can be performed on the lathe.

Ⅱ. Select an answer from the four (or three) choices. *(10×2%)*

1. Which component supports and rotates workpieces about the axis of the lathe?

   (A) Spindle. (B) Carriage. (C) Quill.

2. When advancing the cross slide by .010", the diameter is reduced by ________.

   (A) .005" (B) .010" (C) .020"

3. Which statement is true?

   (A) Turning with hand feed will produce a much smoother and more even finish than is generally achievable by power feed.

   (B) Power feed is a lot more convenient than hand cranking in some applications.

   (C) If carbon steel cutter bits are used, speeds may be 2 or 3 times as high as those given for high-speed steel cutter bits.

4. Which cutter bits allow the greatest cutting speed?

   (A) Carbon steel.

   (B) Carbide-tipped.

   (C) High-speed steel.

5. The device shown in Fig.A1-1 is probably used on a:

   (A) Milling machine

   (B) Turning center

   (C) Machining center

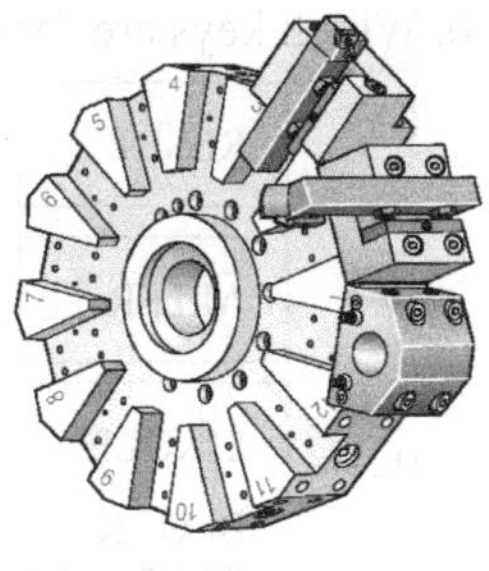

Fig.A1-1 Turret

6. When you rotate the carriage handwheel clockwise of the lathe,

   (A) The tool is moved towards the free end of the work.

   (B) The tool is moved towards the headstock.

   (C) The tool is moved towards the center of the work.

7. On the horizontal milling machine, the horizontal movement of the worktable will be in the:

   (A) Z- and X-axes.

   (B) X- and Y-axes.

   (C) Z- and Y-axes.

8. The spindle speed is usually measured in ________.

   (A) mm/min (B) r/min (C) p/r

9. As shown in Fig.A1-2, this machine can be programmed on ________ axes.

   (A) Three (B) Four (C) Five

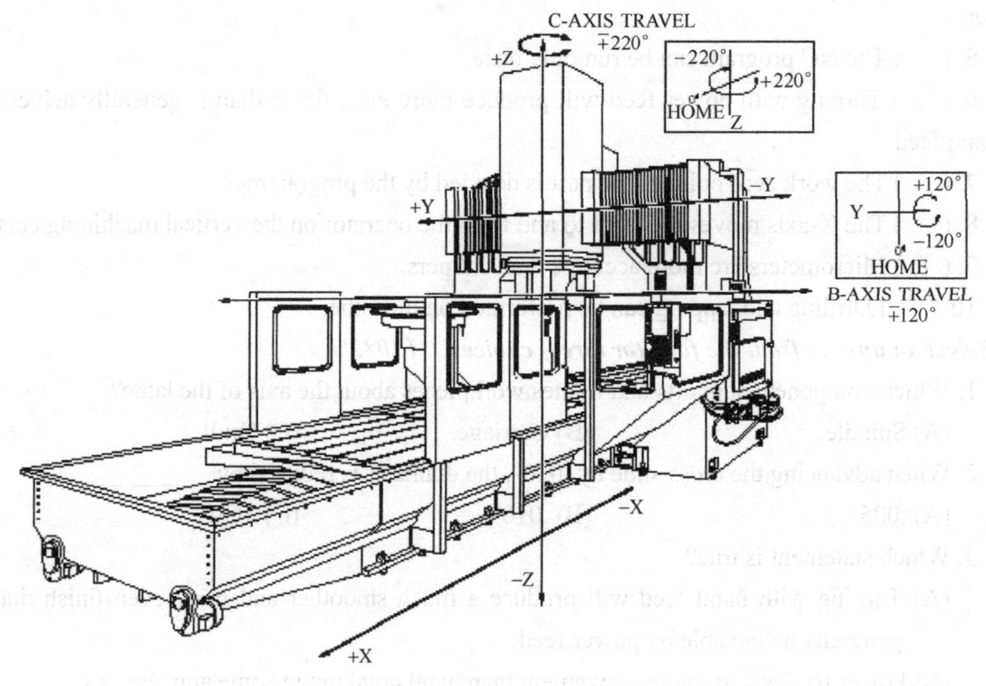

Fig.A1-2 Multi-axis machine

10. Which keys are for editing the NC program in the FANUC series operator's panel?

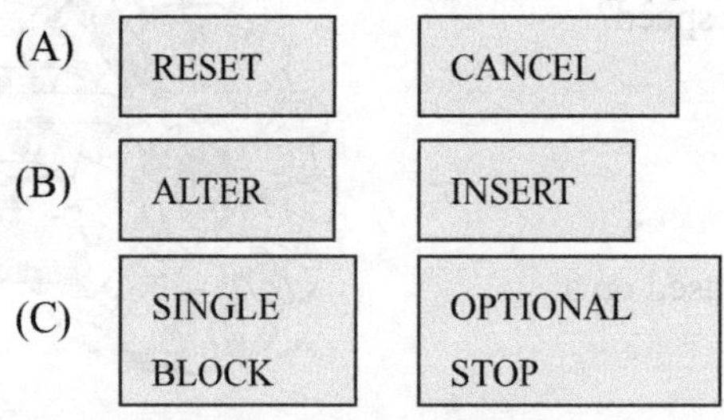

III. Fill in the blanks with the appropriate form of the underlined words. (10×1%)

Example: CNC is a form of programmable automation in which the machine tool is controlled by a program in computer memory.

1. The half-nuts are engaged only when the lathe is used to cut threads, at which time the feed mechanism must be ________.

2. Automobile parts, machine parts and compressors are precision products. They are cut and

shaped by using CNC machines which are extremely ________________.

3. After cutting and finishing operations, a ______________part is formed.
4. Machines that can perform all necessary operations in one setup are called ______________centers.
5. The first benefit offered by CNC machines is improved automation. CNC machine tools are ______________.
6. The second benefit offered by CNC machines is consistent and accurate workpieces. The ____________of the finished part is consistent.
7. The third benefit offered by CNC machines is flexibility. CNC makes manufacturing systems more ______________.
8. These axes can be precisely and automatically positioned along their lengths of travel. CNC controls that utilize a reference point for each axis require that the machine be ________ sent to its reference point in each axis as part of the power up procedure.
9. M03 is used to turn the spindle on in a clockwise manner. M04 turns the spindle on in a __________________manner.
10. Common axis names are X, Y, and Z for linear axes and A, B, and C for __________ axes.

IV. Match the following tools with the correct words using lines. (5×2%)

reamer　　tap　　drill　　endmill　　bore

V. Translate the name of each lathe component into Chinese shown in Fig.A1-3. (10%)

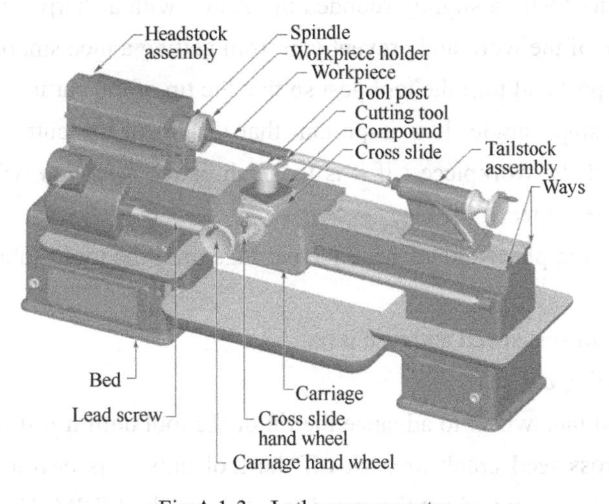

Fig.A 1-3　Lathe components

VI. Translate the following passages into Chinese. (2×10%)

1. The cross slide consists of a dovetailed slide that moves at a right angle to the ways. The compound is mounted on the top of the cross slide. The compound bolts into the disk in the cross slide which enables the compound to be rotated.

2. Writing a program for a numerically controlled tool involves several steps. Before tool programmers can begin writing a program, they must analyze the blueprints of whatever function is to be performed or item is to be made. Programmers then determine the steps that must be taken and what tools will be needed. After all necessary computations have been made, the programmers write the program, most often using computer-aided design (CAD) systems.

VII. Read the following material, and then complete the tasks. (20%)

Facing operations

Facing is the process of removing metal from the end of a workpiece to produce a flat surface. Most often, the workpiece is cylindrical, but using a four-jaw chuck you can face rectangular or odd-shaped（形状奇怪的） work to form cubes（立方形的东西） and other non-cylindrical shapes. Facing is often used to bring the piece to a specified length.

Chucking the workpiece

Clamp the workpiece tightly in the three-jaw chuck. To get the work properly centered, close the jaws until they just touch the surface of the work, then rotate the workpiece by hand in the jaws to seat it; then tighten the jaws. It's good practice to tighten the jaws from all three chuck key positions to ensure even gripping by the jaws. When a lathe cutting tool removes metal it applies considerable lateral（侧面的） force to the workpiece. To safely perform a facing operation the end of the workpiece must be positioned close to the jaws of the chuck. The workpiece should not extend more than 2-3 times its diameter from the chuck jaws unless a steady rest is used to support the free end.

Preparing for the facing cut

Choose a cutting tool with a slightly rounded tip. A tool with a sharp pointed tip will cut little grooves across the face of the work and prevent you from getting a nice smooth surface. Clamp the cutting tool in the tool post and turn the tool post so that the tip of the cutting tool will meet the end of the workpiece at a slight angle. It is important that the tip of the cutting tool be right at the horizontal center line of the workpiece ; if it is too high or too low, you will be left with a little bump（隆起物） at the center of the face.

Clamp the tool post in place and advance the carriage until the tool is about even with the end of the workpiece.

Set the lathe to its lowest speed and turn it on.

Beginning the facing cut

Use the compound handwheel to advance the tip of the tool until it just touches the end of the workpiece. Use the cross feed crank to back off the tool until it is beyond the diameter of the workpiece. Turn the lathe on and adjust the speed to a few hundred RPM. Now slowly advance the cross feed handwheel to move the tool towards the workpiece. When the tool touches the workpiece it should start to remove metal from the end. Continue advancing the tool until it reaches the center of the workpiece and then crank the tool back in the opposite direction (towards you) until it is back past the edge of the workpiece.

Task 1 Translate the first paragraph into Chinese. (10%)

Task 2 How to make facing? Write the steps using the job instruction method based on your experience. (10%)

Assessment 2

I. Decide whether the following statements are true(T) or false(F).　(10×1%)

1. (　) On conventional machine tools, about 80% of the time was spent removing material.
2. (　) Boring and tapping can be performed on the milling machine.
3. (　) The turning center has the same basic axes as the engine lathe.
4. (　) Feed is programmed in either mm/min or mm/r.
5. (　) The turning center is more accurate and more productive than the engine lathe.
6. (　) If the material to be cut is hard and tough, the feed rates are small.
7. (　) A constant cutting speed is important to ensure optimum cutting in a turning operation.
8. (　) A worn or dull tool requires more power to machine a workpiece than a sharp tool.
9. (　) The lathe is widely used in producing box-type work.
10. (　) On a CNC machine, the tool is monitored by man.

II. Select an answer from the four (or three) choices.　(10×2%)

1. Which code will generally determine the measurement system used when machining a part?
   (A) G code.
   (B) S code.
   (C) M code.
   (D) F code.
2. For what industry was the first NC machine used?
   (A) Computer.
   (B) Aviation.
   (C) Automobile.
   (D) Toy.
3. A vertical milling machine has a spindle axis of rotation:
   (A) Vertical with the table.
   (B) Perpendicular to the table.
   (C) Horizontal.
   (D) None of the above.
4. To keep your machine maintained, you must:
   (A) Oil it daily.
   (B) Keep it clean.
   (C) Grease it daily.
   (D) All of the above.
5. In the picture (Fig.A2-1) the upper arrow is pointing to what part of the milling machine?
   (A) Bed.
   (B) DRO.
   (C) Column.
   (D) CNC control.

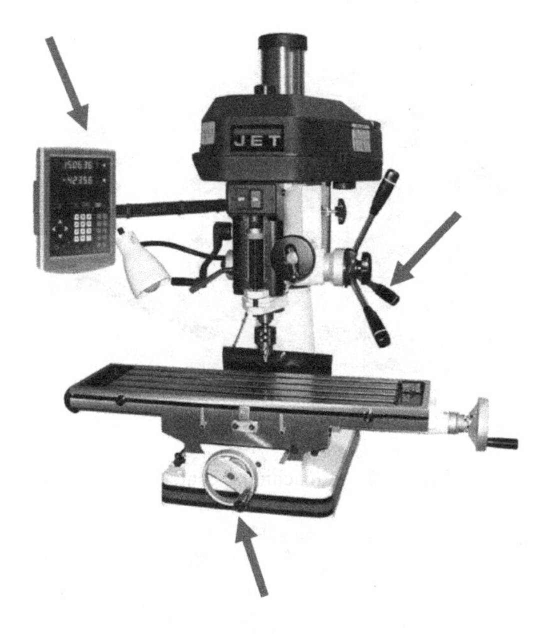

Fig.A2-1　A milling machine

6. In the picture (Fig.A2-1) the middle arrow is pointing to what part of the milling machine?
   (A) Handle.
   (B) Lever.
   (C) Column.
   (D) None of the above.
7. In the picture (Fig.A2-1) the lower arrow is pointing to what part of the milling machine?
   (A) Spindle.
   (B) Column.
   (C) Handle.
   (D) None of the above.
8. In the picture (Fig.A2-2) the arrow is pointing to what part of the machining center?
   (A) Spindle.
   (B) Magazine.
   (C) Tool.
   (D) None of the above.
9. In the picture (Fig.A2-3) the arrow is pointing to what part of the machine control panel?
   (A) Operation modes.
   (B) Alpha keys.
   (C) Overrides.
   (D) None of the above.

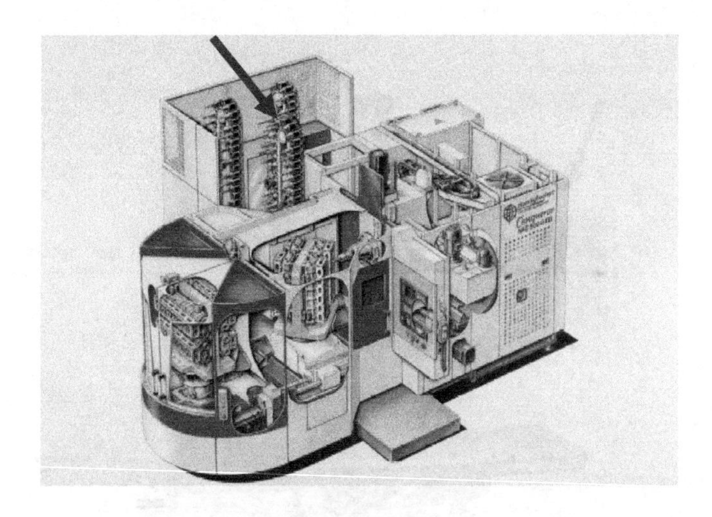

Fig.A2-2 A machining center

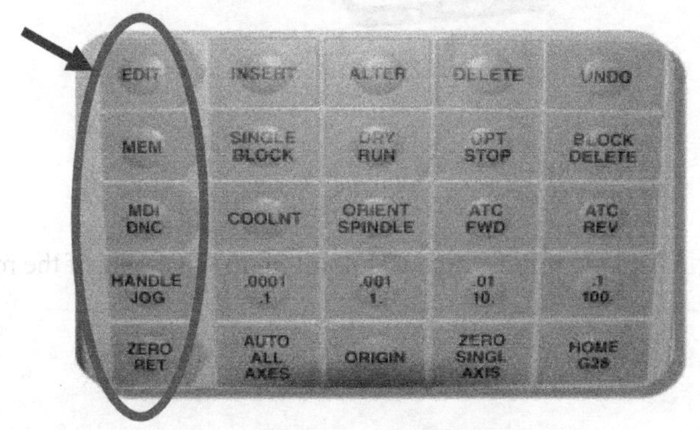

Fig.A2-3 Control panel

10. The operator's panel shown in the picture (Fig.A2-4) is probably made by:

(A) HAAS.

(B) SIEMENS.

(C) FANUC.

(D) None of the above.

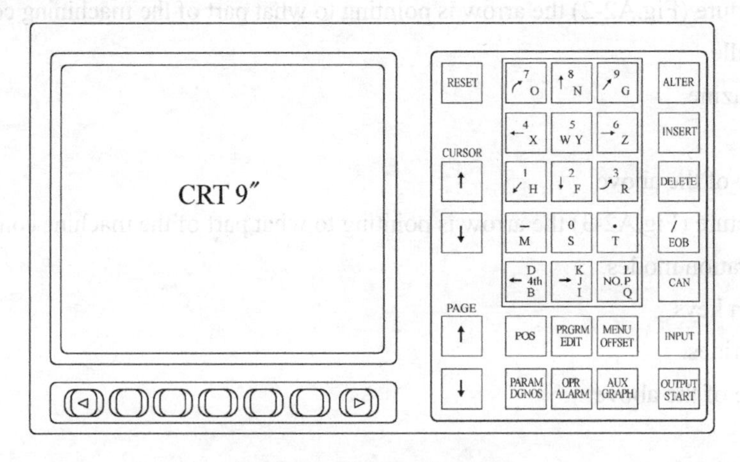

Fig.A2-4 Operator's panel

Ⅲ. Fill in the blanks with proper words. (10×2%)

1. Boring is the operation of producing __________________（良好的表面光洁度） in a hole which has been previously drilled.

2. Speed and feed of threading operations are governed by the____________________（螺纹的导程）.

3. The ____________________ is the main casting mounted on the left end of the bed, in which the spindle is mounted.

4. A numerical control machine is a machine ______________________（自动定位） along a preprogrammed path by means of coded instructions.

5. There are two types of machining centers: _________________________________.

6. As a diameter changes during a machining operation, the spindle speed will automatically increase, decrease, or remain unchanged. This is called ____________.

7. Most CNC machines utilize a very accurate position along axis as a ________________（参考点） for the axis.

8. Newer and better way to assign program zero is through some forms of_____________（偏置）.

9. NC reduces the amount of non-chip-producing time by selecting speeds and feeds, making ______________（快速移动）between surfaces to be cut, using ___________________（自动刀具交换）, and loading and unloading the part.

10. In newer, high speed CNC machines, increasing the cutting speed normally eliminates the ________________（振动）and reduces production time.

Ⅳ. Match the following parts with the correct words using lines. (5×2%)

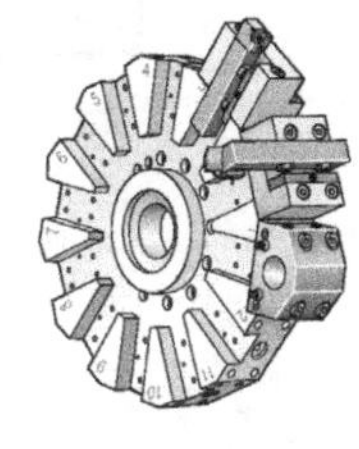

center　　turret　　tailstock　　deep hole drill　　three-jaw chuck

Ⅴ. Translate the following passages into Chinese. (2×10%)

1. The proper cutting speed for a given job depends upon the hardness of the material being worked on, the material of the cutter bit, and the feed and depth of cut to be used. If the cutter bit does not cut satisfactorily, the speed should be reduced. Carbon steel tools, when used, require a reduction in speed because they cannot withstand the heat produced as a result of high speed turning. Carbide tipped tools, however, will stand speeds in excess of those recommended for high-speed steel tools.

2. For a universal style slant bed turning center, for example, the programmer should know the most basic machine components, including the bed, way system, headstock & spindle, turret construction, tailstock, and work holding device.

VI. Read the following material, and then complete the tasks. (20%)

The following material is selected from an MTB operator's manual and explains how the drilling canned cycle is used.

G81 DRILLING CANNED CYCLE

FORMAT:

G81 Z-___ R___ F___

Z = Position of the bottom of the hole being drilled.
R = Reference plane, or a position placed above Z0.
F = Feed rate in inches per minute.

NOTE: The Z, R, and F codes are required data for all canned cycles.

NOTE: The optional X and Y can be included in the canned cycle line. In most cases, this would be the location of the first hole to be drilled.

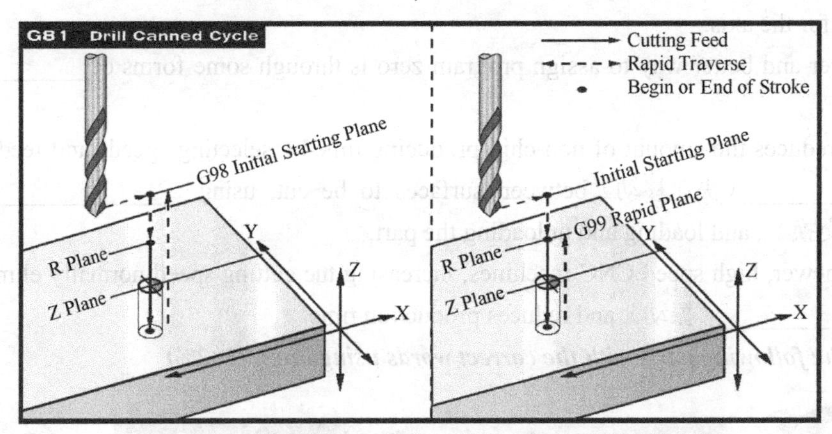

Canned cycle programming example using aluminum block

The following is the program to drill through the aluminum plate:

```
T1  M06
G00  G90 G54  X1.125  Y-1.875  S4500  M03
G43  H01  Z.1
G81  G99  Z-.35  R.1  F27.
X2.0
X3.0  Y-3.0
X4.0  Y-5.625
X5.250  Y-1.375
G80  G00  Z1.0
G28
M30
```

Vocabulary		Notes
reference plane	参考平面	
initial starting plane	起始平面	
aluminum plate	铝板	Z.1 即 Z0.1，Z-.35 即 Z-0.35，R.1 即 R0.1，单位为英寸
optional	可选择的	
drilling cycle	钻削循环	
through hole	通孔	

Task　Examine the following part (Fig.A2-5), and write a program to drill the four through holes using the drilling cycle.

Some data related are:

tool (drill) number : 5

work zero setting: G55 in the lower left corner

spindle speed: 3000r/min

feed rate: 30in/min

【提示】仿照示例，用钻削循环编写图A2-5 中零件的 4 个通孔加工程序。图中所标注的尺寸单位为 mm，先将图中尺寸转换为英寸，再编程。

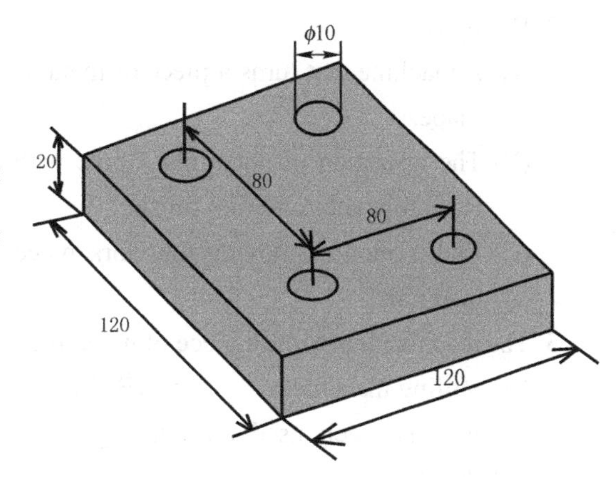

Fig.A2-5　A part

Assessment 3

I. Select an answer from the four (or three) choices.　(10×1%)

1. Which machine is the father of all machine tools?

   (A) The mill.　　(B) The lathe.　　(C) The drill.

2. Boring is:

   (A) A machine that turns a piece of metal round and round against a sharp tool that gives it shape.

   (B) The operation of enlarging a hole with a single-point tool. This operation produces a close tolerance and fine finish.

   (C) The motion of moving the work piece and the cutting tool together so as to remove material.

3. The machine that turns a piece of metal round and round against a sharp tool is:

   (A) Milling machine.　　(B) Lathe.　　(C) Boring machine.

4. The lathe is widely used in producing __________ work.

   (A) Box-type　　(B) Round　　(C) Complex

5. What is the lathe bed made of?

   (A) Aluminum.　　(B) Steel.　　(C) Cast iron.

6. Which component supports and rotates workpieces about the axis of the lathe?

   (A) Spindle.　　(B) Quill.　　(C) Carriage.

7. Milling machine is:

   (A) A machine that turns a piece of metal round and round against a sharp tool that gives it shape.

   (B) A machine that removes metal through the use of electrical sparks（电火花）which burn away the metal.

   (C) A machine tool that removes material by rotating a cutter and moving into the material. It is used to produce flat and angular surfaces, grooves（槽）, contours（轮廓）, and gears.

8. What movement does the table provide the vertical machining center with?

   (A) The Y- and X-axes movement.

   (B) The Z- and Y-axes movement.

   (C) The Z- and X-axes movement.

9. The machining center can automatically select and change the tools that__________.

   (A) Move left and right

   (B) Move up and down

   (C) Have been preset

10. A single NC machine is the same as a machining center in the fact that__________.

    (A) Both of them can select and change tools automatically

    (B) Both of them must be programmed

    (C) Both of them employ a tool magazine

II. Identify each picture with a word or phrase. (10×2%)

	1.	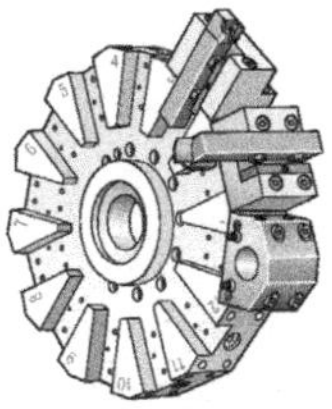	2.
	3.		4.
	5.	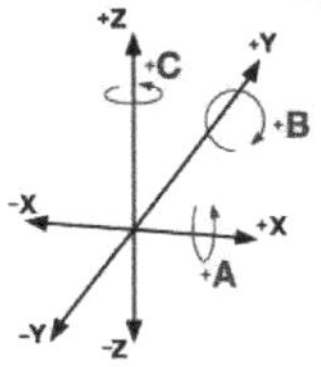	6.
	7.		8.
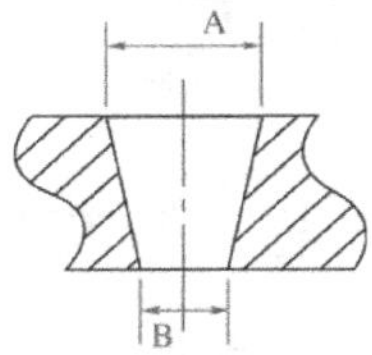	9.	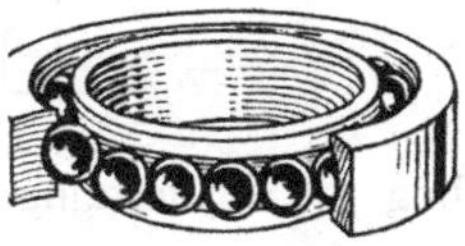	10.

III. You are required to fill the brackets with the appropriate words. (20 × 1%)

1. The turning machine is commonly called __________.

2. The lathe is particularly adapted to cylindrical processing. It may also be used for other purposes, such as ________, ________, ________, ________, ________, __________and so on.

3. The most basic components of the lathe are________, ________, ________, _______ and so on.

4. The main parts of CNC machining centers are bed、column、_______, ________, _______, _______, ________, _________, ________and so on.

5. A horizontal machining center operates on three ______________.

6. Machining centers greatly increased production rates because more operations could be performed on a workpiece in one _______________.

IV. After reading the list of terms, you are required to find the Chinese equivalents in the table below. Then you should fill the brackets with the corresponding letters. (10 × 1%)

A. tool magazine
B. face milling
C. spindle speed
D. CNC drilling center
E. automatic pallet changer
F. machine axes
G. MCU
H. lead screw
I. traveling column type
J. horizontal machining center
K. work holding device
L. spindle range
M. zero return
N. consistent accuracy

Example: (*B*) *平面铣削*

1. (　　) 数控系统	6. (　　) 丝杠
2. (　　) 主轴挡位	7. (　　) 刀库
3. (　　) 钻削中心	8. (　　) 夹具
4. (　　) 回参考点	9. (　　) 卧式加工中心
5. (　　) 动柱式	10. (　　) 自动托盘交换装置

V. Match the following operations with the correct descriptions using lines. (6×2%)

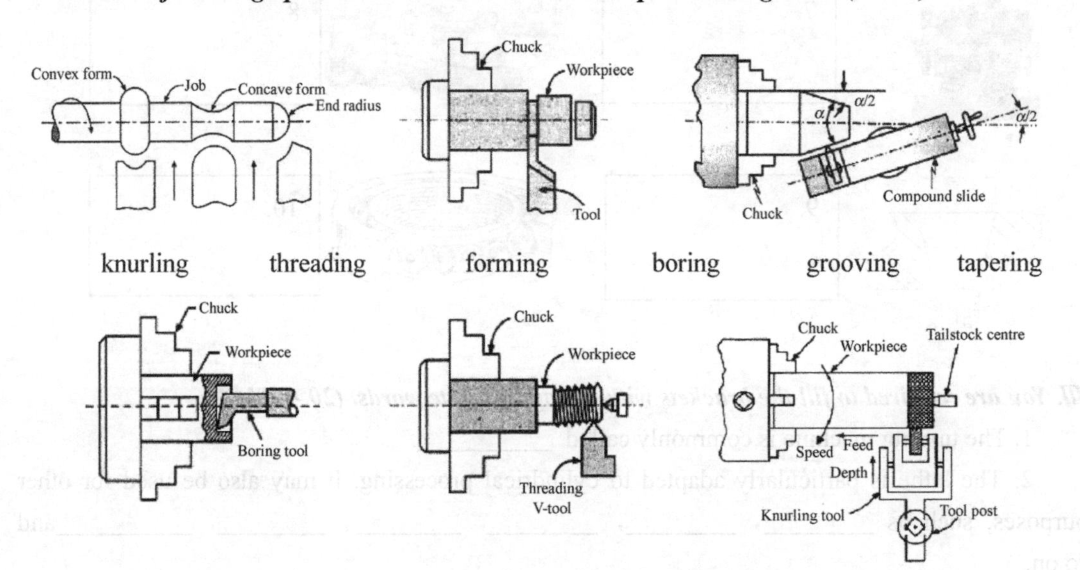

VI. Translate the following technical phrases into Chinese. (10×1%)

1. computer numerical control
2. computer aided programming
3. simultaneous three-axis movement
4. heavy duty machining
5. engine lathe
6. horizontal machining center

7. speed change mechanism

8. contour milling

9. shoulder turning

10. finished product

Ⅶ. Translate the following passages into Chinese. (2×9%)

1. The apron is attached to the front of the carriage. It contains the mechanism that controls the movement of the carriage for longitudinal feed and thread cutting. It controls the transverse movement of the cross slide. In general, a lathe apron contains the following mechanical parts: a carriage handwheel, gear trains, a selective feed lever, friction clutches, half-nut lever and half-nuts.

2. A program is written as a set of instructions given in the order they are to be performed. The instructions might look like this:

LINE #1 = SELECT CUTTING TOOL.

LINE #2 = TURN THE SPINDLE ON AND SELECT THE RPM.

LINE #3 = TURN THE COOLANT ON.

LINE #4 = RAPID TO THE STARTING POSITION OF THE PART.

LINE #5 = CHOOSE THE PROPER FEED RATE AND MAKE THE CUT (S).

LINE #6 = TURN OFF THE SPINDLE AND THE COOLANT.

LINE #7 = RETURN TOOL TO HOLDING POSITION AND SELECT NEXT TOOL.

Assessment 4

Ⅰ. Decide whether the following statements are true(T) or false(F).　(10×1%)

1. (　　) Cutter chatter shortens the life of the cutter.

2. (　　) Most lathes are programmed on three axes: X-axis, Y-axis and Z-axis.

3. (　　) The cutting tools are manually selected on the machining center.

4. (　　) Fine boring is a kind of roughing operation.

5. (　　) Speed and feed of threading operations are governed by the lead of the thread.

6. (　　) As the ratio between length and diameter decreases, the rigidity of the boring bar increases.

7. (　　) A boring bar with a 4:1 length-to-diameter ratio is more rigid than one with 1:1ratio.

8. (　　) A worn or dull tool requires more power to machine a workpiece than a sharp tool.

9. (　　) The actual tool change time on the machining center is usually 10-20s.

10. (　　) Reaming is a kind of roughing operation.

Ⅱ. Select an answer from the four (or three) choices.　(10×2%)

1. In the picture (Fig.A4-1) the arrow is pointing to what part of the FMC?

(A) Panel

(B) Pallet

(C) Turret

(D) None of the above

Fig.A4-1　Flat carrier type FMC

2. We call the switch shown in the picture (Fig.A4-2(a)):

   (A) Override

   (B) EMG stop

   (C) Spindle load meter

3. We call the component shown in the picture (Fig.A4-2(b)):

   (A) Override

   (B) EMG stop

   (C) Spindle load meter

   (D) Handle

4. This switch shown in the picture (Fig.A4-2(c)) is:

   (A) Override

   (B) EMG stop

   (C) Handle

   (D) To select operation mode

5. This button shown in the picture (Fig.A4-2(d)) is:

   (A) Override

   (B) EMG stop

   (C) Spindle load meter

   (D) Handle

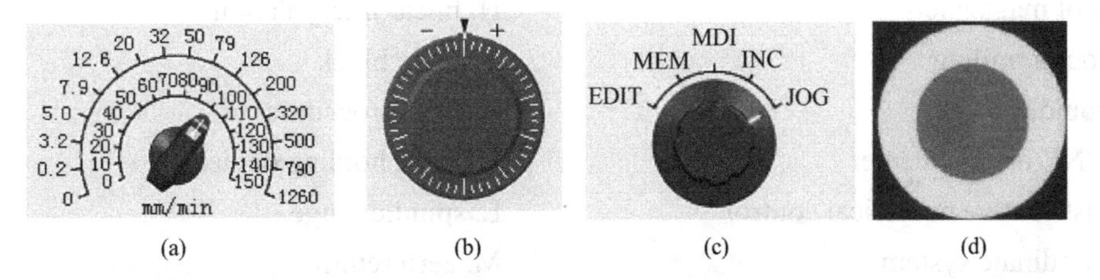

Fig.A4-2 Operating components

6. M01 is usually the code for:

   (A) Program stop

   (B) Optional stop

   (C) End of program

7. For which tools is the length compensation used?

   (A) Drills or taps

   (B) Milling cutters

   (C) Lathe tools

8. G42 will select the cutter radius compensation:

   (A) Left

   (B) Right

   (C) Neither

9. Which command should be used to program the contour according to Fig.A4-3?

(A) G41

(B) G42

(C) G49

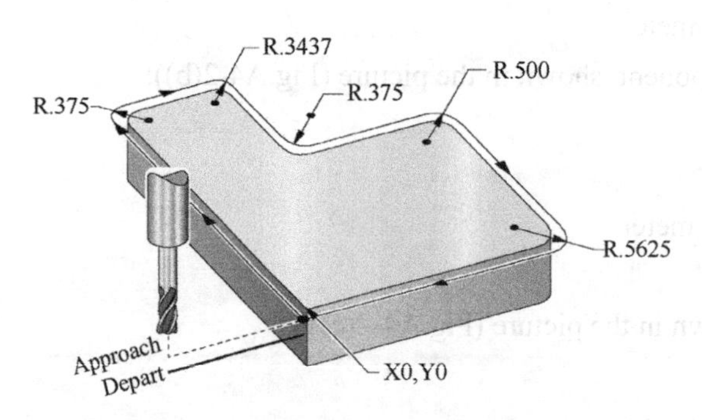

Fig.A4-3　Contour milling

10. The Y-axis on a vertical milling machine:

(A) Controls the table movement left or right.

(B) Controls the table movement toward or away from the column.

(C) Controls the up or down movement of the spindle.

Ⅲ. After reading the list of terms, you are required to find the Chinese equivalents in the table below. Then you should fill the brackets with the corresponding letters. (10×1%)

A. tool magazine　　H. linear interpolation

B. rough milling　　I. single block

C. spindle speed　　J. setup procedure

D. CNC drilling center　　K. work holding device

E. distributive numerical control　　L. spindle range

F. coordinate system　　M. zero return

G. CNC control　　N. consistent accuracy

Example:　(C) 主轴转速

1. (　　) 数控系统	6. (　　) 直线插补
2. (　　) 坐标系	7. (　　) 刀库
3. (　　) 钻削中心	8. (　　) 夹具
4. (　　) 回参考点	9. (　　) 装夹步骤
5. (　　) 单程序段	10. (　　) 粗铣

Ⅳ. Write out the corresponding G code or M code. (10×1%)

1. Spindle Reverse (　　)　　2. Drill Canned Cycle (　　)

3. Coolant Off (　　)　　4. Tool Change (　　)

5. Incremental (　　)　　6. CW Interpolation Motion (　　)

7. Tool Length Compensation+ (　　)　　8. Orient Spindle (　　)

9. Rapid Motion (　　)　　10. Linear Interpolation Motion (　　)

V. Name the following operations shown in Fig.A4-4, using the gerund form of the verb. (5×2%)

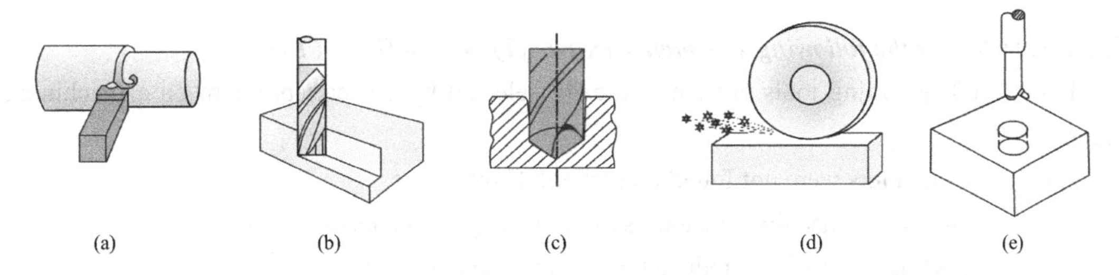

Fig.A4-4 Various operations

VI. Translate the following passages into Chinese. (2×10%)

1. Precision machine line consists of high precision horizontal machining centers and precision grinding machines. With the strict temperature and humidity control, high precision key parts machining is guaranteed by high precision machines and experienced operators.

2. CNC stands for computer numerical control. It allows you to control the motion of tools and parts through computer programs that use numeric data. CNC can be used with nearly any traditional machine. The most common CNC machines found in the machine shop include machining centers (mills) and turning centers (lathes).

VII. Read the following material, and then answer the questions. (20%)

Integrex Series multi-tasking machines represent the full integration of machining centers and turning centers, allowing diverse families of parts（零件族） to be machined in one setup. It's what we call done-in-one. Mazak focused Integrex technology on a portion of our machining operations that manufactures over 80 distinct part numbers, including spindle shafts and chucks for machine tools. Four Integrex machines with 120-tool capacity, 12" chucks, and automatic chuck jaw changers are integrated with a 31-pallet cell that moves workpieces between machines, load/unload stations, and queue stations. Moreover, this 4-machine cell replaces eleven machine tools including three HMCs, two VMCs, two turning centers, four multi-tasking machines, and all of the tooling that went with them.

Palletech palletized（用托盘装运） machining systems are pre-engineered automation that can serve one machining center with a few pallets – or up to eight machines with one hundred pallets. And, customers can add machines, pallets and load/unload stations as needed. Palletech includes a simple, yet effective scheduler, which allows jobs to be queued and prioritized（优先处理） for lights-out operation. Yet when the prototype job or the emergency delivery arises, as it always does, it can be given top priority in the schedule without the need for re-organizing other work.

Questions:

1. What do Integrex Series machines integrate?
2. How many kinds of parts can Integrex Series multi-tasking machines manufacture?
3. What's the use of the pallet according to the material?
4. What is included in the 4-machine cell which replaces 11 machine tools?
5. What is included in Palletech palletized machining systems?

Assessment 5

Ⅰ. Decide whether the following statements are true(T) or false(F).　(10×1%)

1. (　　) The cutting tools are automatically selected by the part program on a machining center.

2. (　　) Computers were not found on earlier NC machines.

3. (　　) On a CNC machine, the tool is monitored by computer-control module.

4. (　　) CNC machines have reduced non-chip-producing time.

5. (　　) G42 is used to specify the tool length compensation.

6. (　　) In this block of program, G01 X100Y250 F200; G01 designates the motion rate.

7. (　　) Tool length compensation command is G43.

8. (　　) In the incremental mode, if a motion mistake is made in one command of the program, only one movement will be incorrect.

9. (　　) G00 shows that the machine will make cuts.

10. (　　) Turning centers allow feed rate to be specified in inches or millimeters per revolution.

Ⅱ. Select an answer from the four (or three) choices.　(10×2%)

1. How often do we clean chips?

(A) Daily.　　　(B) Weekly.　　　(C) Monthly.

2. Which of the following are the two main functions of the cutting fluid in machining?

(A) Improve surface finish on the workpiece and wash away chips.

(B) Reduce forces and power.

(C) Remove heat from the process and reduce friction at the tool-chip interface.

3. What's the proper sequence of changing the processor's board?

① remove a dozen connectors
② reconnect
③ reload
④ download all parameters
⑤ replace the board

(A) ①→④→③→②→⑤

(B) ①→②→③→④→⑤

(C) ④→①→⑤→②→③

4. Tool changing in turning centers is a kind of:

(A) Spindle function

(B) Feed function

(C) Miscellaneous function

5. With the capabilities of maintaining constant surface speed (CSS) at the point of the cutting tool, the spindle speed will automatically _________as the diameter decreases during a turning operation.

(A) Increase

(B) Decrease

(C) Remain unchanged

6. This machine (Fig.A5-1) is:

(A) Horizontal.

(B) Vertical.

(C) Rotational.

7. This operation shown in Fig.A5-2 is called:

(A) Mirror.

(B) Canned cycle.

(C) Interpolation.

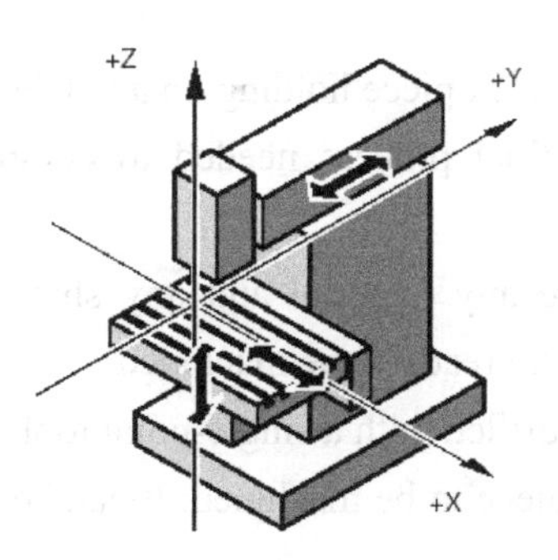

Fig.A5-1 Vertical milling machine

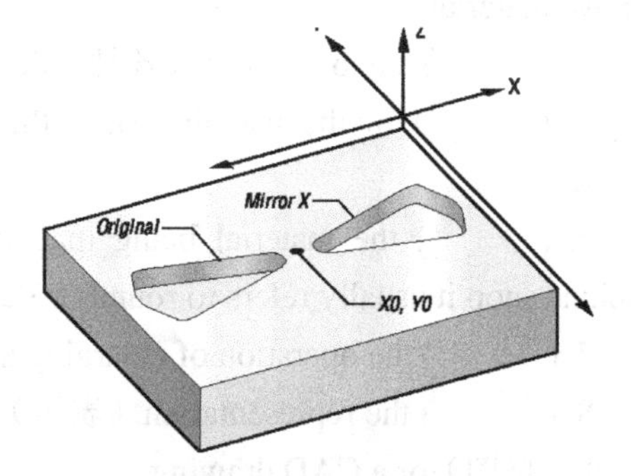

Fig.A5-2 A kind of operation

8. What movement does the column provide the horizontal machining center with?

(A) The X-axis movement.

(B) The Y-axis movement.

(C) The Z-axis movement.

9. Mazak Cyber Production Center is:

(A) A large workshop.

(B) A multi-tasking machining center.

(C) A computer software system.

10. A single NC machine is the same as a machining center in the fact that_________.

(A) Both of them can select and change tools automatically

(B) Both of them must be programmed

(C) Both of them use a tool magazine

Ⅲ. You are required to find the English explanations for the list of terms. Then you should fill the brackets with the corresponding letters. (10×1%)

A. finishing
B. CAM
C. spindle speed
D. absolute
E. stock
F. offset
G. boring
H. blueprint
I. feed
J. program
K. fixture
L. CNC

Example: (L) A form of programmable automation in which the machine tool is controlled by a program in computer memory.

1. ( ) It uses a computer to assist in any or all phases(阶段) of manufacturing（制造）. But more specifically in the machine shop it deals with automatic programming of CNC machinery.
2. ( ) a sequence of coded instructions
3. ( ) the motion of moving the work piece and the cutting tool together so as to remove material
4. ( ) any of various（各种）devices used for work piece holding on a machine tool
5. ( ) In the machine shop this refers to a final process needed to complete the product.
6. ( ) the material being machined. It can be any material and any shape. In the machine shop it usually refers to round or flat pieces of metal ready to be machined.
7. ( ) the operation of enlarging a hole already drilled with a single-point tool
8. ( ) the representation（表示）of the work piece to be machined. It can be a pencil sketch（略图）or a CAD drawing.
9. ( ) In CNC this is a reference to a Cartesian(笛卡儿) coordinate system in which all coordinate locations(位置) are taken (measured) from the origin (原点).
10. ( ) the number of revolutions that the spindle makes in one minute of operation

Ⅳ. After reading the list of terms, you are required to find the Chinese equivalents in the table below. Then you should fill the brackets with the corresponding letters. (10×1%)

A. tool magazine
B. face milling
C. spindle speed
D. CNC drilling center
E. distributive numerical control
F. tool pocket
G. CNC control
H. lead screw error
I. single block
J. setup procedure
K. work holding device
L. spindle range
M. zero return
N. consistent accuracy

Example: (B) *平面铣削*

1. ( ) 数控系统	6. ( ) 丝杠误差
2. ( ) 主轴挡位	7. ( ) 刀库
3. ( ) 钻削中心	8. ( ) 夹具
4. ( ) 回参考点	9. ( ) 装夹步骤
5. ( ) 单程序段	10. ( ) 分布式数控

V. Translate the component names into Chinese shown in Fig.A5-3 on the corresponding number. (10%)

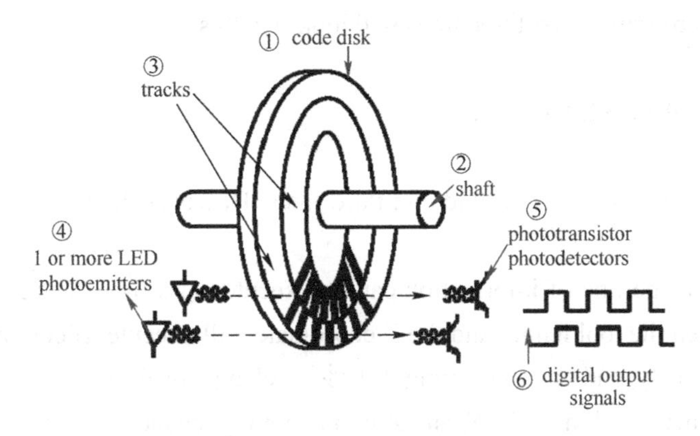

Fig.A5-3 Structure of an incremental encoder

VI. Translate the following passages into Chinese. (2×10%)

1. Many problems are easily overcome by correctly evaluating the situation. All machine operations are composed of a program, tools, and tooling. You must look at all three before blaming one as the fault area. If a bored hole is chattering because of an overextended boring bar, don't expect the machine to correct the fault. Don't suspect machine accuracy if the vise bends the part. Don't claim hole mis-positioning if you don't first center-drill the hole.

2. These machines have three linear axes named X, Y, and Z. The X-axis moves the table left and right, the Y-axis moves it to and from the operator and the Z moves the milling head up and down. The machine zero position is the upper right corner of the mill table. All moves from this point are in a negative machine direction. If a rotary table is connected, an additional A-axis work offset is provided.

VII. Read the following material, and then complete the tasks. (20%)

1. Machines with stepper motors (Fig.A5-4) move in discrete（离散的） length units, since each pulse to the stepper motor makes it rotate through a fixed, finite（有限的） angle. The smallest distance that a machine table can be programmed to move (programming resolution（分辨率） of the machine) is called a Basic Length Unit (BLU).

Example:

A stepping motor of 20 steps per revolution moves a machine table through a lead screw of

0.2 mm pitch.

(a) What is the BLU of the system? (5%)

(b) If the motor receives 2000 pulses per minute, what is the linear velocity in inch/min? (5%)

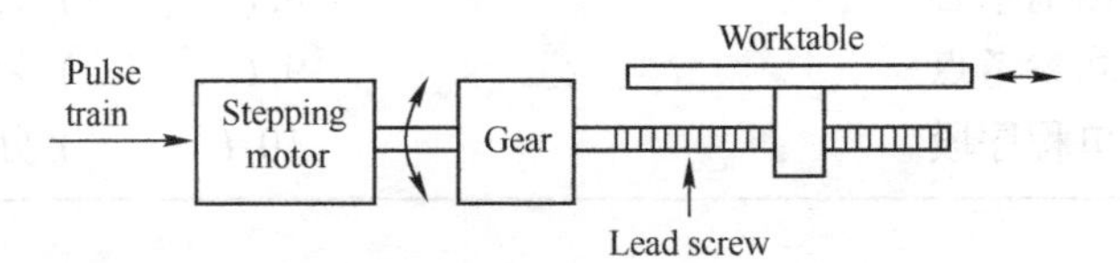

Fig.A5-4 Stepping motor transmission

2. An example program's first three lines will look like this:

T1 M06;

G00 G90 G54 X0 Y0 S2500 M03;

G43 H01 Z.1 ;

We can combine（组合）the second and third lines. However the main reason for using two lines is SAFETY.

Remember, only one line of information can be executed at a time. The X and Y coordinates will position first, then the tool length and the Z coordinate will execute. If combined, all three axes will move simultaneously, and any interfering（干涉） clamps or fixtures（夹具） can be struck and/or destroyed. When combining X, Y, and Z in positioning, chances of crashing the machine are greater. See Fig.A5-5.

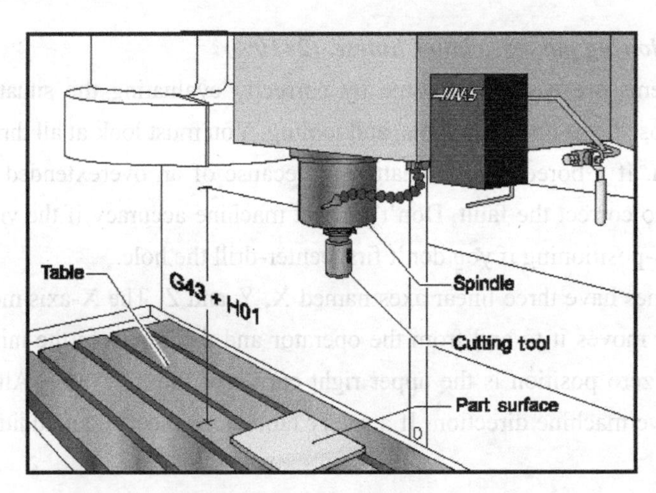

Fig.A5-5 Tool length offset and tool length compensation

Task 1 Write the Chinese name of each part showed in the figure.（5%）

Task 2 We can combine the second and third lines, but why do we write the lines separately? Answer the question in Chinese.（5%）

Appendix A　Suggested translations（参考译文）

学习情境 1　机械工就业

机械工使用诸如车床、铣床和加工中心等机床制造精密金属零件。机械工可能进行一种零件的大批量生产，但精密机械工经常进行单件小批量生产。他们运用其金属材料特性方面的知识和机床方面的技能进行工艺规划和加工，制造满足精度要求的机加工产品。

加工零件之前，机械工必须对整个加工过程进行仔细的规划和准备。首先，这些工人阅读作业零件的图纸或书面说明；接下来，他们计算切入或钻入工件的位置、工件的进给速度、金属的去除量；然后，他们选择适合工件的刀具和材料，制定粗加工和精加工操作的顺序，在金属毛坯上标记切削位置。

规划工作完成以后，机械工就进行必需的加工操作。他们将金属毛坯定位并夹紧到机床（如钻床、车床、铣床或其他机床）上，设定控制方式，然后开始切削。在加工过程中，他们时刻监视机床的进给速度和主轴转速，同时，由于金属件的加工会产生大量的切削热，机械工应确保工件得到充分润滑和冷却。因为大多数金属受热后都会膨胀，所以工件的温度是主要关心的问题；机械工必须基于温度调整切削用量。一些稀有金属的使用日渐增多，如钛金属，它们在超高温条件下进行加工。

机械工通过听特定的声音来判定一些问题，如刀具变钝或剧烈振动时发出的声音。已变钝的刀具会被卸下并更换。调整切削速度以补偿谐振造成的误差，谐振会降低切削精度，尤其在一些新型的高速主轴和车床上。工件加工完后，机械工使用简单和复杂的测量工具，根据图纸要求检查工件的精度。

一些机械工（经常称为制造机械工）可能要大批量地制造某种零件，尤其是那些操作复杂和精度要求高的零件。很多现代机床都是计算机数控机床。机械工经常与数控编程员一起工作，确定自动化的设备如何切削零件。编程员确定切削路径，而机械工确定刀具类型、切削速度和进给速度。由于大多数机械工都接受数控编程的培训，他们会编写基本的程序，试运行过程中，经常修改程序以应对碰到的问题。制造工艺设计完后，相对简单和重复性的操作一般由机床调试工、操作工和放料工完成。

一些制造技术采用自动工件装载装置、自动刀具交换装置和计算机控制装置，使机床能够实现无人化运行。制造机械工一天工作八小时，要监控设备运行，更换用钝的刀具，检查制造零件的精度，同时在几台 24 小时连续运行（无人值守制造）的数控机床上完成其他工作任务。无人值守制造中，一个工厂可能只需要几个机械工就能监控整个工厂的运行。

其他机械工（经常称为机修工）做维修工作——为现有机器修理或制作新零件。为了修理坏了的零件，维修机械工要参考图纸，进行与制造新零件所需的相同的机加工操作。

如今大多数车间比较干净，照明、通风良好，很多数控机床是全防护或半防护的，最大程度上减少了工人暴露在噪声、切屑碎片和工件冷却润滑液中的可能性。不过，在机床周围工作存在一定的危险，工人必须遵守安全预防规定。机械工要身着防护装备，如戴安全眼镜防止飞溅的金属切屑，戴耳套减轻机器噪声。尽管很多常见的以水为基础的润滑液危险性较小，但机械工处理有危险性的冷却液和润滑液时，必须十分小心。机械工的工作需要体力，因为机械工大多数时间都要站着工作，经常还需要搬运重型工件。

由于雇主们看重工人的全面技能，机械工的就业机会持续看好。但是，很多年轻人偏向于读大学，或不愿意涉足制造业。每年都有一些有经验的机械师退休或跳槽到其他职业，因此带来的工作空缺数量大于准备从事机械工工作的工人数量。

所以，不要再犹豫，让我们现在就到车间去看看吧！

学习情境 2　理解普通车床

卧式金属车床常称为车床或普通车床，被认为是机床之父，因为其他机床的设计中都包含了车床的很多基本机械要素。车床应用广泛，可用于加工各种材料。这些刚性的机床通过各种刀具（如车刀和钻头）相对旋转工件进行直线运动去除材料。

要很好地理解车床，需要知道车床的结构和功能。各种部件名称表示在第 22 页的图中。

通常车床由床头箱、床身、床鞍和尾架组成。

床头箱、主轴和卡盘

床头箱由车床左端长方形的金属铸件组成，它包括了主轴及其支撑轴承和变速结构，如图 2-1 所示。床头箱的制造必须十分坚固，因为受切削力的影响可能会使制造不坚固的机座变形，并且产生谐振，谐振传至工件，会降低成品件的质量。

主轴是主要的旋转心轴，卡盘安装其上，即工件随着主轴一起旋转。主轴由安装在床头箱铸件中的精密止推轴承支撑，采用齿轮箱通过电动机驱动。主轴通常是中空的，允许长棒料延伸到工作区，这样就减少了加工准备时间，避免材料的浪费。

位于床头箱前面的主轴转速手柄切换箱内齿轮，起到改变主轴转速和扭矩的作用。离合器控制主轴转向。

卡盘是车床上夹持工件的夹紧装置，车间里常见的有三爪卡盘、四爪卡盘和六爪卡盘。

床身和导轨

床身是车床的主干部分，由坚硬的铸铁制成。导轨是床身上磨削的表面，床鞍和尾架跨骑在床身上，如图 2-2 所示。

床身有倒 V 形、平面式和组合式三种。倒 V 形和组合式床身用于精密轻负荷工件，而平面式床身用于重负荷工件。

床鞍、溜板箱和横向滑鞍

床鞍支撑着中滑板、小滑板和方刀架，在手动或自动进给方式下，沿导轨移动。床鞍由溜板箱和横向滑鞍组成，如图 2-3 所示。

溜板箱连接在床鞍的前方，它包含了控制床鞍的纵向进给和螺纹切削运动的机构，它控制中滑板的横向运动。总之，车床溜板箱包含了下列机械部件：

（a）手动沿床身移动床鞍的大滑板手轮。这个手轮转动齿轮，使之与固定在床身上的齿条啮合。对于短工件的切削，仅使用大滑板手轮进给经常更加方便，但很难获得自动进给下的良好的表面精度。

（b）由光杠驱动的齿轮系。这些齿轮传输来自光杠的动力，从而使床鞍沿导轨方向运动，中滑板穿越导轨横向运动，这样提供自动的纵向和横向进给。

（c）进给选择手柄（这里称为纵向/横向进给手柄）用于按需进行纵向进给或者横向进给的齿轮啮合。

（d）由溜板箱上的纵向/横向进给手柄操作的摩擦离合器用于啮合或断开自动进给机构。有些车床具有独立的离合器，分别用于纵向进给和横向进给，有些则只使用一个离合器，用于两个方向的进给。

（e）开合半开螺母的选择手柄，称之为开合螺母手柄。

（f）车床螺纹切削时用于啮合和断开丝杠的开合螺母。开合螺母由溜板箱上的开合螺母手柄打开或者闭合。开合螺母夹住丝杠时，开合螺母合上丝杠的螺纹，然后丝杠像螺栓一样在螺母中转动。大滑板的移动由丝杠的螺距控制而不是由溜板箱进给机构齿轮控制。只有当车床用于螺纹切削时，开合螺母才啮合，此时进给机构必须断开。通常提供互锁装置作为安全功能，防止开合螺母和进给机构同时啮合。

光杠、丝杠和开合螺母手柄

进给光杠传输动力给溜板箱，驱动纵向和横向进给机构。光杠通过齿轮系由主轴驱动，光杠速度与主轴速度的比率可通过切换齿轮改变，以产生不同的进给速度。旋转的光杠驱动溜板箱内的齿轮，这些齿轮反过来通过摩擦离合器驱动纵向和横向进给机构。光杠用于一般自动切削。

丝杠仅用于螺纹切削，其长度方向为精密切削的 ACME 螺纹，当开合螺母夹住丝杠时，与溜板箱内开合螺母的螺纹啮合，见图 2-4。当丝杠在闭合的开合螺母内转动时，丝杠每转动一周，大滑板沿导轨移动一个螺纹导程的距离。由于丝杠通过齿轮系与主轴相连，丝杠随主轴转动。只要开合螺母啮合，大滑板的纵向运动直接由主轴的转动来控制。主轴每转一转，刀具沿工件移动固定的距离。

待切削螺纹的每英寸螺纹大小和丝杠的每英寸螺纹大小的比率与主轴转速和丝杠转速的比率相同。例如，如果丝杠和主轴选择相同的速度，待切削螺纹的每英寸螺头数与丝杠的每英寸螺头数相同。如果主轴以两倍于丝杠的速度旋转，则待切削螺纹的每英寸螺头数是丝杠的每英寸螺头数的两倍。

只要变换齿轮系中的齿轮获得预期的主轴和丝杠转速比，就可以切削任意螺距的螺纹。

开合螺母手柄通过啮合和断开动力传输来控制丝杠的自动进给。

中滑板和小滑板

中滑板由与导轨成 90° 方向移动的燕尾式滑板组成。小滑板装在中滑板的上面，小滑板通过螺栓与中滑板上的转盘连接，使得小滑板可以旋转。图 2-5 中，小滑板已被卸下，表示了安装在横向滑鞍上的中滑板。

横向进给通过丝杠和手轮向工件以直角（与工件的夹角）进刀，见图 2-6。

小滑板坐在中滑板的上面，可以旋转来设定刀具以某个角度向工件进刀。例如，将小滑板设定在 4°，就可以进行小锥度圆锥切削。有时，将小滑板设定在 30°，这样小滑板手轮前进 0.001 英寸，实际上刀具前进了 0.0005 英寸，因为 sin30°=0.5。

方刀架和刀具

方刀架最多可以装四把刀，每把刀通过六角螺钉锁紧。刀具可以是各种现成的硬质合金刀尖刀具，也可以由用户用常见的刀坯磨成不同刀尖形状的高速钢刀具，见图 2-7。

方刀架绕着小滑板上的大螺栓旋转，通过夹紧手柄锁紧到位。操作者可以按需将方刀架锁在任意角度。

尾架和顶尖

尾架铸件体跨骑在导轨上，沿着床身长度方向移动，以适应不同长度的工件，可以用尾架夹紧螺母将其锁在期望的位置上，见图 2-8。

尾架的主要作用是夹持死顶尖以支撑工件的一端，工件被夹在两顶尖之间加工。但是，尾架也用于夹持活顶尖、锥柄钻头、铰刀和钻夹等。

死顶尖装在尾架套筒的锥孔里，套筒可以在尾架筒内来回移动，用于长度方向的调整。通过手轮移动套筒以进行钻孔或攻丝操作。套筒底部的键槽和键配合，使套筒不能转动。

好了，我们熟悉了车床的基本部件，现在可以开始加工操作了。

学习情境 3　掌握基本的机加工操作

车床常见的操作包括车外圆、车端面、镗孔、钻孔、铰孔、攻丝、切槽、切断、倒角、车锥面、车螺纹、车成形面和滚花，如图 3-1 所示。在做这些操作之前，必须弄懂车床速度、进给量、切削深度这些参数知识。同时，应该遵守良好的安全规程确保安全加工。

三要素

确定某一车削操作最佳的进给量和切削速度取决于被加工材料的类型、刀具类型、工件的直径和长度、期望的切削类型（粗加工还是精加工）、所使用的切削油，以及所使用车床的条件。三要素——切削速度、进给量、切削深度，如图 3-2 所示。

（1）切削速度 V_C

刀具的切削速度定义为一分钟内刀具经过工件表面接触的圆周距离。切削速度（米/分钟）不能与车床主轴转速（转/分）混淆，如图 3-3 所示。为了获得相同的切削速度，对于小直径工件来说，主轴旋转快些，而对于大直径工件来说，主轴旋转慢些。

对于给定工件来说，切削速度取决于工件材料的硬度、刀具的材料、采用的进给量和切削深度。表 3-1 列出了正常情况下外圆车削和螺纹车削切削速度的具体范围。要以这些速度进行加工操作并观察刀具和工件的反应，如果刀具切削并不令人满意，应降低切削速度。使用碳素工具钢刀具时，需要降低切削速度，因为这种刀具不能承受高速车削产生的热量。但是，硬质合金刀具可以承受超出表中推荐的基于高速钢刀具的切削速度。进给量和切削深度应按下文（2）和（3）中描述设置，基于表 3-1 中推荐的切削速度取平均值。如果要增加进给量或切削深度，应按比例降低切削速度以防止过热和过度的刀具磨损。

表 3-1　外圆和螺纹车削时的切削速度

工件材料	外圆车削速度(m/min)	螺纹车削速度(m/min)
铝	60 to 90	15
黄铜	45 to 60	15
软青铜	24 to 30	9
硬青铜	9 to24	6
铸铁	15 to 24	7.5
紫铜	18 to 24	7.5
高碳钢	10.5 to 12	4.5
低碳钢	24 to 30	10.5
中碳钢	18 to 24	7.5
不锈钢	12 to 15	4.5

注：表中速度基于高速钢刀具。如果使用切削油，这些速度可以提高25%~50%。如果使用硬质合金刀具，速度可以高达表中速度的2～3倍。如果使用碳素工具钢，这些速度应降低约25%。

要确定给定切削速度下所需的转速，必须知道待切削工件的直径。计算主轴转速，已知工件直径，可使用下面的公式：

$$S = \frac{1000 \cdot V_C}{\pi \cdot D}$$

其中，S= 主轴转速或 RPM (r/min)；

V_C= 切削速度 (m/min)；

D= 工件直径 (mm)。

（2）进给量 f

进给量是指工件每转一转刀具前进的距离，用英寸/转或毫米/转指定，如图 3-4 所示。

对于粗切削，由于表面不要求很光滑，因此进给量可以相对大些。对大多数材料来说，粗切削的进给量应是 0.010~0.020 英寸/转。对于精切削，由于大进给量下的表面光洁度很差，所以精切削必须使用小的进给量。如果毛坯切削量很大，建议进行一次以上的粗切削，然后以相对高速进行少量精切削。

（3）切削深度 a_p

切削深度调节刀具每次纵向移动工件直径的减小量。刀具每进行一次完整的切削，工件直径减小量为切削深度的两倍。通常，切削深度越深，切削速度就越慢，因为大的切削深度需要更多功率，如图 3-5 所示。

粗加工的切削深度通常是进给量的 5~10 倍，这么做的理由是刀具切削刃与工件接触更多，更利于去除金属。对于进给量为 0.010~0.020 英寸/转的粗加工，其切削深度应在 3/16~1/4 英寸之间。精加工通常切削深度很小，由于切屑很薄，因此可以提高切削速度。

车床安全预防

在加工操作中，操作员必须遵守的顺序是：安全第一，精度第二，速度最后。牢记这些，我们再来看一些车床操作之前和操作当中必须遵守的比较重要的安全预防措施。

- 总是佩戴侧面带防护的安全眼镜。车床可能飞溅尖锐、烫热的金属切屑，非常危险。
- 穿着短袖衬衫，如果着长袖则卷起袖管至肘处。宽松的袖口可能缠进旋转的工件里，很快将手或胳膊卷进一条危险之路。
- 穿工作皮鞋保护脚不被车间地面尖锐的金属切屑和可能掉落的刀具和金属块伤到。
- 解下腕表、项链、颈链和其他首饰。由于戒指有可能被旋转工件钩到，严重伤到手指和手，因此最好也不要戴戒指。
- 将长头发束起来，这样就不会卷进旋转的工件中。想想如果头发被缠住，你的脸会怎样吧！
- 开动车床之前总是再次检查，确保工件已牢固夹紧在卡盘或顶尖中。以低速启动主轴，然后逐渐提高速度。
- 养成用毕立刻取下卡盘扳手的习惯。我们建议只要卡盘扳手在卡盘上，手就不要离开扳手。
- 处理重型卡盘时要小心，安装卡盘时用木块保护车床导轨。
- 操作车床前知道紧急停止按钮的位置。
- 小心处理尖锐的刀具、顶尖和钻头。
- 在不影响操作的情况下，刀具伸出部分越短越好。
- 手指要远离旋转工件和刀具，当金属螺旋状切屑在刀尖形成的时候，千万别去弄断它。工件旋转时千万不要去测量，在调整前，总是先停止车床运转。
- 避免越过旋转的卡盘操作。锉削时，左手持锉刀的柄脚，这样手和胳膊就不会在旋转的卡盘上面。
- 避免机床周围地面有油或油脂，防止滑倒撞到机床上。
- 和人说话之前，要关机。

外圆车削

外圆车削是指从旋转的圆柱工件的外径进行去除的处理。外圆车削用于减小工件的直径，通常减小到指定的尺寸，加工出光滑的表面。工件被车削后，邻近部分经常具有不同的直径。

装夹工件

我们要加工的对象是直径为 1.5 英寸、长度为 6 英寸的 45 号钢。与其直径比起来相对短的工件刚性足够好，我们可以安全地车削它，装夹时只用三爪卡盘而不用支撑工件的自由端，如图 3-6 所示。

对于长一些的工件，其自由端需要车端面打中心孔，用死顶尖或活顶尖通过尾架来支撑。没有这样的支撑，刀具施加到工件上的力将使工件从刀具位置弯曲，结果制造出奇怪的形状。也有可能工件受力而在卡爪里松弛，像一个危险的抛射物一样飞出。

将工件插入三爪卡盘中，拧紧卡爪，旋转工件保证已平稳安装。要求工件尽可能与车床主轴中心线平行。在三个卡盘扳手位置拧紧卡盘保证牢固而均匀的装夹。卡紧之后立刻取下扳手。

调整车刀

选择刀尖带小圆度的刀具，这种类型的刀具能够获得良好的表面光洁度。对于粗切削，如果需要去除大量金属，可以选择刀尖较尖锐的刀具。确保刀具已牢固夹紧在方刀架上。

调整方刀架的角度，使刀具与工件的侧面大致垂直。由于刀具的前部刀刃（副切削刃）被磨出一个角度（副偏角），刀尖的左侧应该切入到工件，而不是整个前部刀刃（副切削刃）与工件接触。

确认开合螺母手柄是不合上的。移动大滑板直到刀尖接近工件的自由端，然后前进中滑板直到刀尖正好接触工件的侧面，向右移动大滑板，直到刀尖正好在工件的自由端外侧。

设定切削速度和进给量

如前面讨论的那样，必须考虑工件的旋转速度和刀具相对工件的移动速度。根据表 3-1 和前面提及的公式，可以方便地确定主轴转速。一般情况下，金属越软，切削速度越快。大多数切削操作可以在几百转/分的速度进行。

手动进给车削

由于这个操作会飞溅出烫热尖锐的螺旋形的金属切屑，因此必须一直佩戴安全眼镜，脸部保持远离工件。

现在前进中滑板手柄约 10 格或 0.01 英寸，逆时针转动大滑板手轮使大滑板慢慢朝床头箱移动。当刀具开始切进金属的时候，保持稳定的手柄转动以获得平稳的切削。手动进给方式难以实现光滑均匀的车削。如图 3-7 所示。

继续朝床头箱移动刀具，直到到达离卡盘卡爪约 1/2 英寸的地方。很明显，刀具不能碰到卡盘卡爪，必须小心操作！

不要移动中滑板或小滑板，顺时针旋转大滑板手轮，将刀具移回工件的自由端。你会注意到刀具在退刀时切掉了少量金属。再前进中滑板 0.01 英寸，重复以上过程，直到对此有比较好的感觉。试着一次前进中滑板 0.02 英寸，可以发现，切削深度加大时，加在大滑板手轮上的力也加大了。

自动进给车削

自动进给车削方式一般可以获得比手动进给方式下光滑均匀得多的表面光洁度。对长工件进行多刀切削时，自动进给也比手动进给方便得多。

自动进给由溜板箱上的纵向/横向进给手柄啮合。启动电动机，现在光杠应该逆时针旋转。将刀具定位于工件末端外侧，前进中滑板 0.01 英寸，此时合上纵向/横向进给手柄，大滑板应在光杠提供的动力下慢慢向左方移动。当刀具到达卡盘 1/2 英寸范围内时，断开手柄，停止大滑板运动，如图 3-8 所示。

现在可以使用大滑板手轮将大滑板摇回起点。如果不将刀具退回一点就返回的话，会发现刀具在工件表面切出一条浅浅的螺旋槽。为了避免这个问题（尤其在精加工时），记下中滑板刻度盘上的刻度值，然后逆时针转动横向进给手柄约半转以退刀。现在手动将大滑板摇回起点，前进中滑板回到刚才记下的刻度位置，再加上 0.01 英寸，重复以上过程。你可以获得良好的、光亮的、光滑的表面光洁度。

正常情况下，需要做一次以上的较大切深的粗切削（0.010～0.030 英寸），随后进行一次以上的较小切深的精切削（0.001～0.002 英寸）。当然，必须要计划好，保证最后的精切削准确地将工件加工到所需直径。

自动进给切削时，必须很小心，刀具不能撞到卡盘。似乎每个人都会发生刀具撞卡盘的事，这将损坏刀具和卡盘，很可能使工件报废。因此必须十分注意，手随时保持在手柄上。

测量直径

大多数时候外圆车削用于将工件直径减小到指定尺寸。在外圆车削过程中，每次进刀的金属去除量是中滑板进给刻度指示的两倍，认识到这一点很重要。这是因为你减少的是根据刻度盘指示的工件半径，对直径来说，就减少了两倍的刻度指示量。因此，当中滑板进刀 0.010 英寸时，直径就减少了 0.020 英寸。

工件直径用卡尺或外径千分尺测量。外径千分尺更精确，但不够通用。需要机械师常用的能够精确至 0.001 英寸的卡尺。游标卡尺没有表盘，读数时需要在刻度尺上插入一些数值，而带表盘的卡尺则能提供直接易读而不易出错的测量，如图 3-9 所示。

很明显，工件在运动时千万不要去测量。车床停止后，用带表卡尺的卡爪卡住工件，用球形旋钮收紧卡爪。至少测量两次以确信测量是正确的，这是比较好的做法。

学习情境 4　认识数控机床

数控机床用户必须了解所用数控机床的基本工作原理和结构。

基本的机械加工实践——用好数控机床的关键

很多数控机床可以增强或替代普通机床的功能。数控初学者的第一个目标是理解数控机床使用过程中所需的基本机械加工实践，初学者对基本的机械加工实践懂得越多，就越容易适应数控加工工作。

这样想想吧！因为是否懂得机械加工与将来能否用好数控机床关系很大，所以，如果你已经懂得基本的机械加工实践，你就知道想要机床做什么。这就是为什么机械师能造就最优秀的数控编程员、操作员和调试员的原因。机械师已经知道机床要做的事，将已知的东西运用到数控机床也是相对简单的事。

例如，初学数控车削中心（如第 73 页图所示）者应理解有关车削操作如粗精车、粗精镗、割槽、螺纹切削、凹槽加工等基本的加工实践。由于这种数控机床能在一个程序中完成多种加工（很多数控机床都能够如此），初学者同时应该懂得如何加工车削类零件的基础知识，这样才能编制出待加工零件的加工工序。

运动控制——CNC 的核心

任何数控机床最基本的功能都是具有自动、精确、一致的运动控制。大多数普通机床完全运用机械装置实现其所需的运动，而数控机床是以一种全新的方式控制机床的运动。所有的数控设备都有两个或多个运动方向，称为轴。这些轴沿着其长度方向精确、自动定位。最常用的两类轴是直线轴（沿直线轨迹）和旋转轴（沿圆形轨迹）。

普通机床须通过旋转摇柄和手轮产生运动，而数控机床通过编程指令产生运动。通常，几乎所有的数控机床的运动类型（快速定位、直线插补和圆弧插补）、移动轴、移动距离以及移动速度（进给速度）都是可编程的。

数控系统中的 CNC 指令命令驱动电动机旋转某一精确的转数，驱动电动机的旋转随即使滚珠丝杠旋转，滚珠丝杠将旋转运动转换成直线轴（滑台）运动。滑台上的反馈装置（直线光栅尺）使数控系统确认指令转数已完成，如图 4-1 所示。

普通的台虎钳上有着同样的基本直线运动，尽管这是相当原始的类比。旋转虎钳摇柄就是旋转丝杠，丝杠带动虎钳钳口移动。与台虎钳相比，数控机床的直线轴是非常精确的，轴的驱动电动机的转数精确控制直线轴的移动距离。

学习新数控机床的要点

从编程员的角度看，在开始学习任何一台新的数控机床的时候，应该重点学习四个基本方面的内容。首先，应该理解机床的基本组成部分；第二，应该对机床的轴运动方向很熟悉；第三，应该熟悉机床所配的附件；第四，应该查明机床包括的所有可编程功能，并学会如何对其编程。

机床部件

最常见的数控机床是车削中心和加工中心。车削中心用于加工圆形零件，加工中心用于加工平面或斜面零件。第 73 页图表示了这两种机床。例如，对于通用斜床身式车削中心（见图 4-2），编程员应该知道最基本的机床部件，包括床身、导轨系统、床头箱和主轴、转塔刀架结构、尾架以及工件夹紧装置。机床制造厂提供的说明书中通常有包括装配图纸在内的有关机床结构的信息。阅读机床厂家的说明书时，对下面一些有关机床规格和结构的问题，应该能找到答案。

- 机床的最大转速是多少？
- 机床有几个主轴挡位（各换挡速度是多少）？
- 主轴和进给轴驱动电动机的功率各是多少？
- 各轴的最大行程是多少？
- 机床能容纳几把刀具？
- 机床的导轨结构如何（通常是方形导轨、楔形导轨和/或直线导轨）？
- 机床的快速移动速度是多少？
- 机床的最大切削进给速度是多少？

新数控机床的使用初期，用户应该提出这几个问题。事实上，对机床的规格和结构了解得越多，就越容易适应机床。

运动（轴）方向

数控编程员必须知道数控机床的可编程运动方向（轴）。轴的名称随不同的机床类型而不同，总是以字母地址表示。常用的轴名称是：用于直线轴的 X、Y、Z、U、V、W 和用于旋转轴的 A、B、C，这些轴与坐标系相关，如图 4-3 所示。但是，并非所有的机床制造厂家使用的坐标系都与上述坐标系的名称一致，因此初学编程者应根据机床厂家的说明书确认这些轴的定义和方向（正和负）。

各轴参考点

大多数数控机床使用各坐标轴上的一个精确位置作为各轴的起点或参考点。一些数控系统制造商把这个位置称为回零位置，有些厂家则称它为栅格位置，还有人称之为原点位置。不管叫什么名称，参考位置是许多数控系统所要求的，目的是给数控系统一个精确的基准点。各轴采用参考点的数控系统需要手动将机床的各轴返回参考点，这是上电过程中必不可少的一部分。一旦回参考点完毕，数控系统将与数控机床的位置同步。

机床附件

数控初学者应该关注的第三个方面是与机床的附件有关的内容。很多数控机床配有附件，用以增强数控机床的功能。其中一些附件由数控机床制造商制造和提供，这些附件应在机床说明书中详细记载。其他的附件由产品售后制造商提供，此时会包含一本单独的有关附件的说明书。

数控机床的附件包括探头测量系统、刀具长度测量装置、后处理测量系统、自动托盘交换装置、自适应控制系统、车削中心棒料自动输送装置、车削中心复合动力刀具和 C 轴、自动化系统。其实，附件装置处于不断的发展之中。

可编程功能

编程员也必须知道数控机床的可编程功能。对于一些低价位的数控设备，很多机床功能多数时候必须用手动的方式激活。比如，对于一些数控铣床，唯一的可编程功能就是轴运动，其他的功能都得由操作员激活。对于这种机床，主轴转速和转向、冷却液及换刀都得由操作员手动激活（开启）。

另一方面，对多功能的数控设备，几乎所有的功能都是可编程的，操作员只需装卸工件。一旦循环开始，操作员可空出手来做其他事情。

用户可以参考机床厂家的说明书，查一查所用机床的可编程功能。为了用例子说明有多少可编程功能要处理，这里列出一些常用的可编程功能及相关的编程字（指令代码）。

主轴控制

“S”字用于指定主轴转速（加工中心上用转/分钟或 RPM 作为计量单位），M03 用于主轴顺时针（正）旋转，M04 用于主轴逆时针（反）旋转，M05 使主轴停转。注意车削中心还有恒线速度功能，恒线速度功能就是允许主轴速度用英尺/分钟（或米/分钟）指定。

自动换刀装置（加工中心）

“T”字用于告诉机床哪把刀要装到主轴上。在多数机床上，M06 命令机床进行换刀。4 位“T”字用于命令大多数车削中心上的换刀。前两位数字指定刀架的刀位号，后两位数字指定刀具的偏置号。例如，T0101 指定了 1 号刀位和 1 号偏置。

冷却液控制

M08 用于开启水状冷却液。如果有 M07，则用于开启雾状冷却液。M09 用于关闭冷却液。

自动托盘交换装置

如图 4-4 所示，M60 指令通常用于托盘交换。

其他可编程功能

如前所述，各机床的可编程功能很不一样，实际所需的编程指令也随制造厂家的不同而不同。一定要检查机床说明书中的 M 代码表（辅助功能，参见附录 E），多了解所用机床的其他可编程功能。M 功能通常由机床厂家指定，用于给机床用户提供可编程的开/关功能。无论如何，必须知道数控程序中可激活的功能。

例如，对车削中心来说，你会发现尾架和尾架套筒是可编程的，卡盘卡爪的开/合是可编程的。如果机床有两挡以上的主轴挡位，通常主轴挡位的选择是可编程的。如果机床有送棒料装置，它也是可编程的，甚至可以发现机床的排屑装置可通过编程指令进行开和关。当然，所有这些对数控编程员来说都是重要的信息。

学习情境 5　学习 HAAS 控制面板

HAAS 数控系统介绍

HAAS 数控系统是目前车间里最常用的数控系统之一，尽管别的数控铣床会采用不同制造厂家的机床本体和数控系统，可每一台 HAAS 铣床都配置 HAAS 数控系统。详细学习了数控系统后，就会发现 HAAS 数控系统设计合理、使用方便。

控制面板

由控制面板控制所有的机床操作。控制面板（见图 5-1）由手动控制部件、显示屏和控制键盘组成：

- POWER ON（电源开）和 POWER OFF（电源关）按钮用于开机和关机。
- SPINDLE LOAD（主轴负载）表显示主轴电动机的功率输出。它告诉操作员机床的工作负载，防止机床因过载而损坏。
- EMERGENGY STOP（紧急停止）按钮自动关闭所有的机床功能。如果眼看刀具就要与工件或夹具相撞时，就应该使用该按钮。
- HANDLE（手轮）使机床部件沿着坐标轴移动。手轮转动时发出滴答声，可通过键盘选择移动增量。
- CYCLE START（循环启动）按钮开始执行程序或重新启动中断的程序，FEED HOLD（进给保持）按钮使轴运动停止，但主轴运动继续运行。要停止主轴，按 SPINDLE STOP（主轴停止）修调键。FEED HOLD 用于检查工件或刀具的加工情况。
- 显示屏显示程序、轴的位置，以及整个加工过程中其他相关的信息。
- 控制键盘用于输入命令、输入偏置、调整速度。

图 5-2 和图 5-3 表示 HAAS 铣床的手动控制部件。HANDLE（手轮），EMERGENCY STOP（紧停），CYCLE START（循环启动）和 FEED HOLD（进给保持）这些手动控制部件的功能很像其他机床上的手动控制部件。

控制键盘

控制键盘是 HAAS 数控系统输入命令的地方。如图 5-4 所示，控制键盘分 9 个区：

- 8 个显示键位于控制键盘的顶部中间位置。
- 30 个操作方式键位于控制键盘的右上角。
- 15 个数字键位于控制键盘的右下角。
- 30 个字母键位于控制键盘的底部中间位置。
- 15 个修调键位于控制键盘的左下角。

- 8 个光标键位于控制键盘的中心位置。
- 15 个手动进给键位于修调键上方区域。
- 8 个功能键位于手动进给键的上方。
- 3 个复位键位于控制键盘的左上角。

显示键——顶行

显示键位于键盘的顶部中间位置。按下各键显示不同的信息，如偏置、机床设置和当前正在运行的程序。其中一些键连按两下就显示另外一个页面和其他信息。显示屏的左上角显示当前信息。顶行显示键（见图 5-5）提供了成功运行程序所必需的信息：

- 按 PRGRM/CONVRS（程序/对话）显示当前程序的程序段。按两下 PRGRM/CONVRS 用于进入编辑方式下的对话编程页面。
- 按 POSIT（位置）显示 5 个不同页面，列出机床轴的位置。
- 按 OFFSET（偏置）显示各种偏置。
- 按 CURNT COMDS（当前命令）显示运行过程中 15 行当前程序、模态指令及轴位置。

根据加工任务不同，可能需要选择特定的显示页面。装入新刀并输入偏置时，可以选择 OFFSET 页面。运行零件程序时，选择 PRGRM/CONVRS 或 CURNT COMDS 页面。

显示键——底行

底行显示键（见图 5-6）提供了附加的显示和信息。如顶行显示键一样，底行中的一些显示键如果被按两下，就显示另外一个页面。

- 按一下 ALARM/MESGS（报警/信息）显示生效的报警。在报警显示下，按光标右移键，显示报警历史，而按光标右移键两下，显示报警说明。按 ALARM/MESGS 两下，显示一条信息，提供给当前操作员或下一位操作员。
- 按一下 PARAM/DGNOS（参数/诊断）列出机床参数。机床参数很少改变。HAAS 公司建议用户在无法确定需要改什么和为什么要改，以及没有正确咨询车间和 HAAS 服务人员之前，不要随意更改参数。如果参数被随意更改，则用户可能无法得到机床的保修。按 PARAM/DGNOS 两下显示诊断页面。
- 按一下 SETNG/GRAPH（设定/图形模拟）显示设定页面，用于设定机床控制功能，用户可能需要激活、撤销或更改这些机床控制功能，以符合具体的工况。本课后面将学习具体的设定。按两下 SETNG/GRAPH，用户可以从屏幕上看到零件程序的运行情况，从而避免由于编程错误造成刀具和机床损坏的危险。
- 按一下 HELP/CALC（帮助/计算）显示信息说明，包含了 26 个帮助主题。按两下 HELP/CALC 可以选择三角函数计算器、圆弧数据计算器、铣削/攻丝切削数据计算器、圆弧-直线相切计算器或圆弧-圆弧相切计算器。

操作方式

如图 5-1 所示，操作方式键位于控制键盘的右上角。操作方式告诉控制系统要完成的任务，而且一次只能处于一种操作方式，可以从图 5-7 中左边列出的按钮中选择具体的操作方式。选择某一操作方式后，与方式键在同一行的其他键就生效了。如图 5-8 所示，当前操作方式显示在显示屏的顶行，就在当前显示页面的右侧。有 6 种操作方式：

- EDIT（编辑）方式用于手动修改程序或建立新程序。可以 INSERT（插入）、ALTER（替换）、DELETE（删除）或 UNDO（撤销）程序。
- MEM（存储器）方式从数控系统的存储器运行零件程序。按 MEM 显示当前程序，当按下 CYCLE START（循环启动）时，程序将开始执行。
- MDI/DNC（手动数据输入/直接数字控制）方式使机床处于手动数据输入方式。MDI 用于在不干扰存储程序的情况下，输入并执行程序。按 MDI/DNC 两下，如果 Setting 55（设定值 55）接通，则激活直接数字控制。DNC 用于执行来自软驱或计算机硬盘的程序。
- HANDLE/JOG 方式允许操作员用手轮或手动按钮移动各坐标轴。
- ZERO RET（回零）方式用于寻找机床零点或自动快速返回机床零点。
- LIST PROG（程序列表）方式列出程序，用于选择、发送、接收以及删除程序。必须在 PRGRM/CONVRS 显示页面和 LIST PROG 方式下建立新程序。从键盘上输入 Onnnnn，再按下 SELECT PROG 按钮。按下 EDIT 显示新程序。

字母键和数字键

数字键（见图 5-9）位于控制键盘的右下角，用于输入数字(0～9)、负号(–)和小数点(.)。

字母键（见图 5-10）显示了按字母表（A～Z）排列的所有大写字母。小写字母（a～z）可在括号之间输入，只要按住 SHIFT 键，然后输入字母即可。

特殊字符

图 5-11 和图 5-12 显示了各种特殊字符，这些特殊字符在程序中具有特定的含义。

修调键

修调（倍率）键（见图 5-13）位于控制键盘的左下角，用于改变程序中编写（简称程编）的进给速度和主轴转速、主轴转向和快速移动速度。通过使用修调键，用户可以在程序运行时调整这些变量。

每按一下–10 FEED RATE（进给速度）当前进给速度以 10%递减，每按一下+10 FEED RATE 当前进给速度以 10%递增。100% FEED RATE 使进给速度回到程编进给速度。HANDLE CONTROL FEED（手轮控制进给速度）用于旋转手轮以 1%的增量调整程编速度，此时，设定值 104 Jog Handle to Single Block 为 OFF。

每按一下–10 SPINDLE（主轴转速）当前主轴速度以 10%递减，每按一下+10 SPINDLE 当前主轴速度以 10%递增。100% SPINDLE 使主轴速度回到程编值。HANDLE CONTROL SPINDLE（手轮控制主轴转速）用于旋转手轮以 1%的增量控制主轴转速，此时，设定值 104 为 OFF。

最后两行，CW 启动主轴顺时针方向旋转，STOP 使主轴停止转动，CCW 启动主轴逆时针方向旋转。5%RAPID、25%RAPID 和 50%RAPID 用于按给定的百分比减小快速移动速度，100% RAPID 使机床以最快的速度运行。

光标键

如图 5-14 所示，光标箭头键也用于编辑和搜索数控程序。用箭头键可以在程序或页面选项中上、下、左、右移动光标。

点动键

点动键（见图 5-15）位于光标键左侧区域。如果按住点动键，被选中的轴将连续运动。+X 和–X 按钮使 X 轴运动，+Y 和–Y 按钮使 Y 轴运动，+Z 和–Z 按钮使 Z 轴运动。同时按 SHIFT 和+A 或–A，能使 B 轴运动。在操作上述键之前按下 JOG LOCK 键，则不需要一直按着轴进给键，轴即能连续运动。再次按 JOG LOCK 键将停止轴运动。

点动键左侧是控制螺旋排屑器的按钮。CHIP FWD（排屑正转）使螺旋排屑器转动起来，从机床上排除铁屑，CHIP STOP 停止排屑器的运动，CHIP REV 使排屑器反转。

点动键右侧是控制机床冷却液的按钮。如果冷却喷嘴有效，CLNT UP（冷却液向上）使冷却水流方向上升一个位置，CLNT DOWN 使冷却水流方向下降一个位置，如果在 MDI 方式，AUX CLNT 将开启主轴中心孔冷却系统。按 AUX CLNT 两次，关闭该系统。

功能键

8 个功能键（见图 5-16）位于点动键的上方。功能键能完成很多不同的任务。有一些功能键只完成一种任务，另外一些则取决于当前显示的页面。

顶行功能键 F1, F2, F3, F4 根据不同的显示和操作方式完成不同的功能。这些键就位于 RESET（复位）键的下方，和编辑、图形模拟、后台编辑、帮助/计算器一起，用于执行特殊功能。

底行功能键包括执行特定任务的 4 个键：

- TOOL OFSET MESUR（刀具偏置测量）用于零件装夹时在偏置页面中（自动）设定当前 Z 轴刀具长度偏置值。
- NEXT TOOL（下一把刀）用于零件装夹时激活换刀装置，选择下一把刀。
- TOOL RELEASE（刀具松开）用于 MDI、手轮/点动或回零方式下从主轴中松开刀具。松刀按钮在刀具松开之前必须按 1.5 秒，按钮松开后，刀具将保持松开状态 1.5 秒。
- PART ZERO SET（工件零点设置）用于零件装夹时自动输入工件坐标偏置值。

复位键

复位键位于控制键盘的左上角，图 5-17 表示了这些键的具体位置。按红色的 RESET 按钮，可停止所有机床运动，将程序指针放到当前程序的开头。

POWER UP/RESTART（上电/重新启动）按钮在上电时将机床初始化。上电之后，该键可用于重新初始化系统。

TOOL CHANGER RESTORE（换刀装置恢复）按钮用于换刀过程被中断时，将换刀装置恢复到正常状态。这个键调出一个用户提示页面，帮助操作员进行换刀碰撞故障恢复。

基本操作

熟悉了数控系统和控制键盘后，就可以给机床通电，甚至可以进行一些运行零件循环前的基本操作。

上电过程

HAAS 铣床的通电过程分几个简单步骤完成，如图 5-18 所示。

1．按下标有 POWER ON（电源开）的绿色按钮。该按钮位于控制面板的左上角，如图 5-19 所示。

2．按下如图 5-17 所示控制键盘上的 POWER UP/RESTART 按钮，使各轴运动至机床零点参考位置。图 5-20 表示了铣床上的机床零点。

机床零点由机床制造厂设定并且固定不变，它是机床坐标轴正向的极限点。在机床进行任何操作之前，必须先找到其固定的机床参考点。

记住：在 POWER ON 按钮接通铣床之前，机床后侧的主断路器必须接通。任何电源中断将使铣床停机，POWER ON 按钮用于再次接通电源。

回零方式

除了 POWER UP/RESTART 键能使机床回到零点位置，还有其他几种方法可使机床回零。

机床通电后，在用 POWER UP/RESTART 键回零之前，不知道其原位（Home），也可以先按 ZERO RET（回零）方式键，再按 AUTO ALL AXES（自动将所有轴回零）键，从而使机床回到零点。这些键如图 5-21 所示。这使各轴回到原位，并且将所有轴都初始化回到机床零点。

如果只需一个轴初始化归零，则只要在 ZERO RET 方式下按 ZERO SINGLE AXIS（单轴回零）。这将对输入缓冲区中用字母指定的轴进行初始化。

最后，ZERO RET 方式下，HOME G28 键可用于在不进行初始化的条件下，使机床各轴快速返回机床零点。由于没有警告信号提醒用户可能发生碰撞，因此必须特别小心，以保证机床回参考点时不会与其他物体发生干涉。如果有可能碰撞，按下 EMERGENCY STOP（紧急停止）按钮。

激活冷却液

冷却液可减小刀具和工件之间的摩擦和磨损，由于冷却液是制造过程的重要组成部分，检查冷却液面，知道如何激活冷却液非常重要。

要检查老版本 HAAS 铣床上的冷却液面，只需用量油计。在新版本 HAAS 数控系统上，可从 CURNT COMDS 页面上观察冷却液传感器，如图 5-22 所示。不管是什么版本的控制系统，确保充足的冷却液非常重要。

设定页面可控制操作过程中冷却液的使用。按 SETNG/GRAPH 键观察设定表。Setting 32（设定值 32）控制冷却泵。

- 在“Normal（正常）”设定下，M08 和 M88 冷却液命令照常执行。
- 在“Ignore（忽略）”设定下，程序中的 M08 或 M88 命令不能开启冷却液，可通过 COOLNT（冷却液）键手动开启冷却液。
- 在“Off（关闭）”设定下，冷却液绝对不可以开启，数控系统读到程序中的 M08 或 M88 命令时，发出报警。

程序运行时可以按如图 5-23 所示的 COOLNT（冷却液）按钮手动开启或关闭冷却泵。这会改变当前程序的运行状态，一直保持到执行另一个 M08 或 M09 冷却液指令。

同时不要忘了 Jog 键右侧列有控制机床冷却液的按钮，如图 5-15 所示。如果冷却液喷嘴有效，CLNT UP 键使冷却水流方向上升一个位置，CLNT DOWN 键使冷却水流方向下降一个位置，如果处于 MDI 方式，AUX CLNT 将开启主轴中心孔冷却系统。按 AUX CLNT 两次，关闭该系统。

激活排屑装置

机加工操作产生切屑，切屑是金属加工的副产品。从切削区排除切屑非常重要。如果切屑不及时排除，将影响机床的切削能力。可开启排屑器排除切屑，传送装置将切屑送到小车里，小车用于将切屑运到车间的切屑集中处理区。

控制切屑运动的手动键如图 5-15 所示。程序运行时，可用手动键或程序中的 M 代码开启或关闭排屑器。CHIP FWD（排屑正转）使螺旋排屑器转动，从而从机床上排除铁屑。CHIP STOP 停止排屑器的运动。CHIP REV 使排屑器反转。

M 代码控制机床的辅助功能。与 CHIP FWD 对应的 M 代码是 M31，CHIP REV 是 M32，CHIP STOP 是 M33。在新型的 HAAS 控制系统上，M32 不再用于控制排屑器的反转。

留言

关机或换班之前，按两下 ALARM/MESGS（报警/信息）按钮，可以给下一位操作员或给操作员自己输入信息。ALARM/MESGS 键如图 5-24 所示。可就有关机床状态或保养注意事项留言。机床通电时，自动显示所留信息。

有时，机床需要通知操作员其一般操作状态，如果按 ALARM/MESGS 键一下，机床显示所有当前报警，报警由序号和说明两部分组成。可用 RESET 键一次解除一个报警。按左、右光标键可以显示数控系统已识别出的最近 100 条报警的历史，历史记录包括报警的日期和时间。

关机

有几种方法可以将 HAAS 铣床关机。最常用的方法是按红色的 POWER OFF 按钮，立即切断电源。机床执行程序时，不可以用 POWER OFF 关机。还有两种设定（SETNG 页面）可以使机床在特定条件下关机。

- Auto Power Off Timer（自动关机定时器），或 Setting 1（设定值 1），是指示机床自动关机前的空闲时间的数字设定值。程序运行或操作按键时定时无效。
- Power Off At M30（M30 时自动关机），或 Setting 2（设定值 2），设置为 ON 时，M30 代码结束程序时自动使机床断电。

以上两种设定有效时，数控系统在关机前会给操作员 15 秒来决定是继续运行还是关机。如果决定继续运行，则按下任意键将中断定时器的倒计时。同时，这些设定会首先自动将 Setting 81（设定值 81）Tool at Power Down（关机时的刀具）中所列的刀具装入主轴。

其他情况可能导致机床停机。报警 176 中显示的持续过电压或报警 177 中显示的过热情况，也会使机床自动关机。在机床停机前各种情况必须持续 4.5 分钟以上。

总结

HAAS 数控系统是如今车间里最常用的数控系统之一，HAAS 铣床控制面板有 3 个不同的区域。首先，HAAS 铣床的手动控制，例如 HANDLE, EMERGENCY STOP, FEED HOLD 键，作用如同其他机床的手动控制。第二，在加工过程中显示屏显示所有相关的信息。第三，键盘允许操作员输入机床命令。

HAAS 铣床控制键盘有 9 个区域，每个区域是一组具有类似功能和作用的键。控制键盘由显示键、操作方式键、数字键、字母键、修调键、光标箭头键、点动键、功能键和复位键组成。

操作员了解了控制面板和键盘，就能完成各种基本工作，包括开机和关机、检查和激活冷却液、使机床回参考点、激活排屑器及留言。

学习情境6　机床维护和故障诊断

卧式加工中心维护

以下是卧式加工中心（见图 6-1）的常规维护事项，有维护频率、所需油液容量与类型。为了使机床工作正常，保护保修权利，用户必须遵守这些必要的规范。

每日保养

- 每 8 小时应加满冷却液（尤其在大量使用主轴中心孔冷却液时）。
- 检查导轨润滑油箱液位。
- 清理导轨防护罩和底板上的铁屑。
- 清理换刀装置上的铁屑。
- 用干净的布毯清洁主轴锥孔，并涂上清油。

每周保养

- 检查过滤调节器自动排出口是否正常工作，如图 6-2 所示。
- 对于带主轴中心孔冷却（TSC）的机床，清理冷却箱上的铁屑收集篮。
- 卸下冷却箱盖子，清除其中的沉淀物。注意要关闭冷却液泵，处理冷却箱前要切断数控系统电源。对无 TSC 的机床每月做一次这个工作。
- 检查气压表/调节器为 85 磅/平方英寸（0.6MPa）。
- 根据机床规格检查液压平衡压力。
- 涂少量油脂于换刀装置机械手的外边沿，并对全部刀具都用机械手换一遍。

每月保养

- 检查齿轮箱中的油位。将油加到油开始从废油罐底部的溢流管滴出为止。
- 清理托盘底部的衬垫。
- 清理 A 轴和上料工位上的定位垫。此项操作须卸下托盘。
- 检查导轨防护罩是否正常运行，必要时用清油润滑。

每半年

- 更换冷却液，彻底清洗冷却箱。
- 检查所有软管和润滑管路是否破裂。

每一年

- 更换齿轮箱润滑油。从齿轮箱中将油排尽，慢慢注入 2 夸脱（1.9 L）的美孚 DTE 25 润滑油。
- 检查润滑油过滤器，清除过滤器底部的残余物。

润滑表如表 6-1 所示。

每两年

● 更换控制箱上的空气过滤器。

表 6-1 润滑表

系统	导轨润滑和气动系统	机械传动	冷却液箱
位置	机床右侧控制面板下方	主轴箱后部	机床侧面
说明	活塞泵 30 分钟循环工作一次；只有当主轴旋转或进给轴移动时泵才工作		
润滑部位	直线导轨和滚珠丝杠螺母	仅传动系统	
数量	2～2.5 夸脱，取决于泵循环	2 夸脱	80 加仑
润滑液	Mobile Vactra#2	Mobile DTE 25	仅水溶性冷却液

故障诊断

本节旨在确定一个已知问题的解决方案。所提供的解决方法用于给数控机床维修人员一个可遵循的模式，首先，确定问题的根源，其次，解决问题。

利用常识

通过正确判断当时的故障情况，许多问题很容易解决。所有机床操作都是由程序、刀具和加工方法组成的。在认定一个故障之前，首先看看这三项工作情况。

如果由于镗杆过长使得镗刀振动，不要指望机床修复这种故障。如果虎钳将工件压弯，不要怀疑机床精度。如果不打中心孔，不要断定是孔定位不准。

首先发现问题

很多机械师在问题清楚之前总是指责这不好那不好，事实上，所有包修返回零件中有一半以上工作良好。如果主轴不转，要记住主轴与齿轮箱相连，齿轮箱与主轴电动机相连，主轴电动机由主轴驱动装置控制，主轴驱动装置与 I/O 板相连，I/O 板由 MOCON 模块控制，MOCON 由中央处理器控制。意思是说，如果皮带断了，不要更换主轴驱动装置。发现问题是第一位的，不要只更换最易操作的部件。

不要胡乱修改机床

机床上有成百上千个参数、电线、开关等可以改变。不要随意更换部件和修改参数。记住，如果在维修过程中修改参数、线路等，很可能安装出错或破坏其他元件。考虑一下更换处理器板吧！首先，必须下载所有参数，卸下一大堆插头，更换线路板，重新连接，重新装载参数，如果出错或弄弯一个小插脚，系统便不工作了。在机床上维修时，总要考虑损坏机床的风险，决定更换之前应对可疑件再次检查，这是一种廉价的保险措施。在机床上进行的维修越少越好。

学习情境 7 理解自动化工厂

自动化工厂是指从原材料进入到成品出厂极少需要或不需要人的干预。这就需要计算机辅助设计/制造系统（CAD/CAM）对机器人、工业机器人和数控机床进行编程，以将原材料变成成品，还需要测量机器人完成自动检测。自动化工厂是集成了信息技术和制造技术的完全计算机化的工厂。

无人值守制造是指工厂完全自动化，现场不需有人。因此，这些工厂不需开灯就可以运行。很多工厂能够实现无人值守生产，但是很少有完全灭灯运行的。典型情况是，需要工人装夹待制造的零件，卸下完工的零件。随着无人值守制造技术的不断成熟，很多工厂开始在轮班之间（或单班）采用无人值守制造，以满足日益增长的需求或节省成本。

马扎克是全球知名的机床制造商品牌。小巨人机床有限公司（LGMazak）是马扎克在中国的制造工厂，成立于 2000 年 5 月，坐落于宁夏省。这里我们来看一看 LGMazak 是如何实现工厂自动化和无人值守制造的——马扎克智能网络工厂。

高效率的生产得益于公司采用了全新的制造概念，即加工过程柔性化、数字化、精密化，管理过程网络化、信息化、智能化。公司的销售、生产、技术、财务等的计算机网络化管理使 LGMazak 成为中国第一座智能网络工厂，如图 7-1 所示。

车间生产线

恒温车间建有大件加工线、中小壳体类零件加工线、轴类及盘类零件加工线、精密加工线、部件装配线、涂装作业线、总装作业线、全自动立体仓库、精密检测室。车间内建有恒温超净室，用于机床主轴部件等精密部件的装配及检验。先进的生产设备和厂房使 Mazak 成为世界一流的机床制造公司。

四台典型龙门式五面加工中心组成的大件加工线用于床身、立柱、滑座、工作台和其他大型工件的加工。所有粗加工和精加工由一次装夹完成，保证了零件的精确定位，为后期高精密高效率装配提供了保障。

壳体类零件加工线由三条柔性制造线 FMS 组成。操作者只需要完成 FMS 生产计划的定制，整个 FMS 线就可以按照既定计划进行无人化连续运转。加工零件的精度依靠高精度设备来保证，可消除操作人员人为因素对加工精度的影响。

轴类、盘类零件加工线由八台 Integrex 系列机床组成。Integrex 系列车铣复合中心完全改变了传统的加工概念。工件一次装夹，可以完成从原材料到成品件的全部工序的加工，零件的加工精度可以得到充分保证，成就了高效、高精度的加工。

精密加工线由高精度卧式加工中心和精密磨床组成。在温度和湿度得到严格控制的环境下，通过高精度设备和具有专业经验的操作人员保证关键零部件的精密加工。

钣金生产线由两台激光切割机 FMS 和六台精密数控液压折弯机组成，用于机床防护罩、冷却液箱等钣金零件的加工。

生产支持软件

LGMazak 的每一台机床都配置了 MAZATROL 系列数控系统和柔性制造系统，通过系统的网络化功能将设备连接到公司内的局域网，并通过智能生产中心（CPC，生产支持软件）来控制，从而获得高效生产。如图 7-2 所示。

通过智能生产中心的 CAMWARE 模块，MAZATROL 加工程序可由计算机生成。CAMWARE 能够利用在线生产的刀具、夹具等信息生成加工程序文件，使得编程周期大大缩短。通过 CPC 的智能刀具管理模块，刀具管理人员可以监控在线机床的刀具使用状态，准备备用刀具，及时更换超出寿命期的磨损刀具。生产管理人员可以通过智能生产中心的智能日程管理模块为每台机床定制准确的生产计划，操作者可以直接通过机床查看自己的生产计划，并反馈计划的执行进度。规划管理人员可以查看工厂的生产情况，需要时进行任务分配。操作人员可以通过智能生产中心的智能监控模块实时查看每一台机床的运行情况，并随时进行调整，提高效率。实时反馈，构建看得见的生产体系，通过整个局域网加以控制，加上柔性化的生产设备就可以确保满足客户的交货期要求。

学习情境 8　参加机床展览会

展馆外……

李：欢迎来到本届中国国际机床展览会。现在正在举行开幕仪式。

布朗：没错，场面很热烈。

李：看那些在空中飘扬的大气球，上面写着欢迎标语。

布朗：真的很感人，看起来是一个大展会。

李：一点也不错。已经登记的参观者已达 45,000 人，来自于世界各地，这个数目在未来两三天会继续增加。

布朗：能有这次机会，我感到很荣幸。

李：你说得对。本次展览会对中国的机床工业非常重要。有将近 1,000 台新机床参展，代表各个国家和地区的展团参加了本届 CIMT。

布朗：太棒了！你们每年都有这样的展会吗？

李：实际上，自从 1989 年第一届中国国际机床展览会以来，每两年在中国举行一次。该机床展览会被公认为世界机床领域四大展销活动之一。

布朗：下次我肯定会来的。

李：欢迎，欢迎！

展馆内……

布朗：看！这些展品真的很壮观！

李：当然了！这是 8A 展馆，由四部分组成。这里展出的是一些国产新机床，其中很多已经赶上国外同类产品的技术水平。我带你转转。

布朗：你真是太好了！哟，那儿是一台大型五轴联动数控机床。

李：是的。这台新机床达到了世界先进水平，适用于航空工业。

布朗：哦，是这样。

布朗：这是镗铣床吗？

李：对，由于它经济实用、操作简便、性能优良，世界上很畅销。

布朗：有意思，但是机床尺寸小了点。

李：它是专为小工件设计的，稳定性好，效率高。大一点的也有，你看，在那儿。

布朗：单价多少？

李：这是我们的价格表，这是产品样本。

在谈判间……

布朗：我们已经看过你们的样本，对你们的镗铣床很感兴趣。但通过计算发现价格高了一点。

李：你们要订多少？

布朗：如果价格合适，我们计划订 15 台。

李：说实话，由于供不应求，这种机床已销售一空了。现在价格表上的价格和以前是一样的。但是，由于你们的订单比较大，同时我们想与你们建立长期的合作关系，我们准备把单价降到上海离岸价 20,000 美元，那是我们的底价。

布朗：看起来可以接受。什么时候交货？

李：三个月，8 月份可交货。

布朗：我们希望在今年 10 月份使用这些机床。时间非常紧迫，由于从上海到拉各斯没有直达航运，我们要把机床运到新加坡中转，7 月中旬能交货吗？

李：7 月可以。

布朗：我们什么时候签合同？

李：明天下午。

布朗：明天见！

李：明天见！

Appendix B　Oral English patterns

打招呼

Greetings	Responses
Hi. /Hello. 你好！	Hi. /Hello. 你好！
How do you do? 你好！	How do you do? 你好！
Good morning/afternoon. 早上/下午好！	Good morning/afternoon. 早上/下午好！
How are you recently? 最近怎么样? Long time no see! 好久不见! How's it going? 怎么样？ How are things going? 事情进展得怎样? How's everything? 一切还好吧? How have you been? 你近来过得怎么样？What's going on with you? 你怎么样？ How are you doing? 你好吗？	Fine, thanks. How about you? 挺好，谢谢！你呢? Pretty good, thank you. And you? 相当好，谢谢！你呢? So far, so good.　目前还不错。 Not bad.　还不错。

介绍

Introductions	Responses
Introducing yourself: Hello, I'm (full name). 你好！我是…… Let me introduce myself. I'm (full name). 让我自我介绍一下，我是…… My name is (full name). 我叫……	Nice/ Good/ Glad/ Pleased/ Delighted to meet you. 见到你很高兴！ How nice to see you. 见到你真好！ It's a pleasure to meet you. 见到你很高兴！ Very glad/happy to meet you. 很高兴见到你！
Introducing another person: Let me introduce (name). 让我介绍…… I'd like you to meet (name). 我让你见一下…… I'd like you to meet my classmate (name).我让你见一下我的同班同学…… (Name), this is (name). ……，这是…… Allow me to introduce (name). 允许我介绍…… May I introduce you to (name)? 我可以向你介绍……吗？ Have you met Diana (before)? 你（以前）见过戴安娜吗？	How do you do? 你好！

谈天气

Talking about the weather	Responses
Fine weather, isn't it? 天气很好。	Yes, it's lovely. 是的，很好。
It's quite cold today. 今天很冷。	Really. 真的很冷。

讨论

Opinions	Responses
In my opinion… 我的观点是…… I think… 我认为…… I feel… 我觉得……	Absolutely. 绝对是这样。 I couldn't agree with you more. 我非常同意你的观点。 I agree. 我同意。 I am on your side. 我支持你。
What do you think? 你怎么认为？	Yes，I suppose so. 是的，我也这么认为。 I think so. 我也这么想。

称赞

Compliments	Responses
Fabulous! 好极了！ Great! 太好了！ Terrific! 太棒了！ Fantastic! 太了不起了！ Wonderful! 太精彩了！ Marvelous! 太不可思议了！ Good job! 做得好! You did fairly well! 你干得相当不错！	Really? 真的吗？ Oh, thank you. 谢谢！ It was nice of you to say so. 你这么说真好。
I really like your presentation. 我真的非常喜欢你的报告。	Thank you for saying so. 谢谢你能这么说。 That's very kind of you. 你真是太好了！

感谢

Gratitude	Responses
I really appreciate your help. 我真的非常感谢你的帮助。 It's very kind of you. 你真太好了！ Thank you so much. 非常感谢！ You've been a big help. 你对我的帮助很大。 I can't express how grateful I am. 我不知如何感谢你。	Don't mention it. 不用谢。 You're welcome. 不客气。 Not at all. 没什么。 That's all right. 没关系。 My pleasure. 很乐意。

帮助

Help	Responses
Offering help: Can I help you? 我能帮你吗? What happened to you? 你怎么了？ What's up? 有什么事吗? What's wrong with you? 你哪里不对劲?	Oh, yes please. 好吧！ No, thank you. 不用，谢谢。 I come across a problem. 我碰到一个问题。 I have trouble with my machine. 我的机床出故障了。
Requests: Can you do me a favor? 帮个忙，好吗? Excuse me，Sir. 先生，对不起。 Give me a hand! 帮帮我! Could you…, please? 你能…… ？ You might help me with… 你可以帮我……	Yes, of course. 当然可以。 I'd be glad to. 我很乐意。 OK. /Fine. 好的。 Sure. 当然了。 My pleasure. 我很乐意。 If you like. 只要你愿意。

鼓励

Encouragement	Responses
Never mind. 不要紧。 Take it easy. 别紧张。 I'm on your side. 我全力支持你。 Keep on going. 继续。 Never give up. 不要放弃。	Thank you. 谢谢。 I will try. 我会努力。 You are so kind. 你真好。

请客

Invitation	Responses
Let me treat you. 我来请客。 You are my guest. 我来请客。 It is your/my treat. 你/我请客。 My treat. 我请客。 I'd like to invite you to dinner this Saturday. 我想邀请你这星期六吃饭。	Thank you, I'd love to come. 谢谢，我很乐意。 Sorry, I can't. But thank you anyway. 抱歉，我没法来。但是谢谢你的好意。

说再见

Goodbye	Responses
Goodbye！再见！ I should go now. 我该走了。	Bye-bye! 再见！
See you (later). 再见。	See you (later). 再见。
It's nice meeting you. 很高兴认识你。	It's nice meeting you. 很高兴认识你。
Keep in touch. 保持联系。	Keep in touch. 保持联系。
Take care. 保重！	That's very kind of you. 你真是太好了！
I'm looking forward to seeing you soon. 期待不久后再见到您。 I hope to see you again sometime. 我希望下次再见到您。	Me too. 我也是。
Have a pleasant journey. 祝您旅途愉快！	Thank you. 谢谢！

Appendix C Technical vocabulary

2D display	二维显示	air gauge	气压表
3D display	三维显示	air pressure	气压
4-pole	4 极	aircraft	飞机
5-axis machining	5 轴加工	alarm	报警
		allow	允许
A		allow for	考虑到，顾及
A-axis	A 轴	alloy steel	合金钢
absolute	绝对的	alphanumeric character	字母数字字符
absolute encoder	绝对编码器	aluminum	铝
absolute programming	绝对编程	alter	替换
abrasive	研磨剂	alternate	交替的
acceleration[反]deceleration	加速（度）	amount	数量
access (to)	存取	analog	模拟的
access protection	存取保护	A/D	模数转换
accessory	附件	angle	角度
accompany	伴随	angle encoder	角度编码器
accomplish	完成	anneal	退火
accuracy[近]precision	精度	anode[反]cathode	阳极
achieve	获得	aperture	窄缝
acknowledge	应答	apparatus	仪器，装置
ACME thread	梯形螺纹	apply (to)	将……应用于
activate[反] deactivate	激活	application	应用
active	活跃的，生效的	appliance	用具，器具
actual	实际的	apprenticeship	学徒的身份
actual value	实际值	approach	接近，进刀；方法
adapt (to)	与……适应	apron	溜板箱
adaptive control	适应控制	approval	批准，许可
additional	附加的	approximate	大约的，接近的
add-on	附加，附件	approximity switch	接近开关，无触点开关
address	地址；从事（*v.*）	arbor	刀柄，心轴
adjust[近]tune	调整	arc	圆弧，电弧
advance	前进，进步	architecture	建筑
advantage	优势，优点	arithmetic	算术的
aerospace	航空	armature	电枢；衔铁
aerospace part	航空零件	armor plate	钢板
aid	帮助，辅助	around the clock	连续不断地
air cooled	空气冷却的，气冷的	as a rule of thumb	凭经验

assembly 装配
assembly drawing 装配图
assembly line 流水线
assignment 分配，指定，作业
associated (with) 与……相关的
assume 承担；假定
asynchronous motor 异步电机
attain 获得，达到
auger 螺丝钻
authority 权威，威信
auto tool changer (ATC) 自动刀具交换装置
automate（*v.*） 使自动化
automatic（*a.*） 自动的
automatically（*ad.*） 自动地
automation（*n.*） 自动化
automobile 汽车
automotive 汽车的
automatic pallet changer(APC) 自动托盘交换装置
auxiliary 辅助的
auxiliary axis 辅助轴
auxiliary function 辅助功能
available 可提供的，有的
average 平均
axis 坐标轴

B

B-axis B 轴
back grounding edit 后台编辑
backlash 反向间隙
backlash compensation 反向间隙补偿
backup （数据）备份
backup battery 备份电池
backspace 退格键
ball screw 滚珠丝杠
band saw 带锯
bar stock 棒料
bar code 条形码
bar feeder 送棒料装置
batch 批量，一批
battery 电池
Baud rate 波特率
beam sag compensation 横梁下垂补偿
bearing 轴承
bed way 床身导轨
belt 皮带
bend 弯曲
bending machine 折弯机
bevel 斜面
bias 偏置，偏移
bidirectional 双向
binary 二进制
bit （二进制的）位
blade 刀刃，刀片
block 程序段
block search 程序段检索
block delete 程序段删除
blueprint 蓝图
board 板
bolt 螺钉
bond 键，接头
boost 提升
bore 镗（孔）
boring machine 镗床
boring-milling machine 镗铣床
boring tool 镗刀
boring bar 镗杆
brake 制动
branch office 分公司
breaker[近]air switch 空气开关
brittleness 脆性，脆度
brochure 小册子
brush 电刷
brushless 无刷的
buffer 缓冲器
built-in 内置的
bus 总线
bush 衬套，轴瓦
byte 字节

C

C-axis	C 轴
cable	电缆
cabling	连线
calculation	计算
caliper	卡钳
call	调用
cam	凸轮
cancel	取消
canned cycle	固定循环
capable of	能做…
capability	能力，功能
capacity	容量
capital	大写；资本
capture	俘获
carbide	硬质合金
carbide insert mill	硬质合金镶嵌式铣刀
carbon tool steel	碳素工具钢
card	卡
carousel	环形传送装置
carriage	车架；托板
carry out	进行，完成
Cartesian coordinates	笛卡儿坐标
cartridge	夹头，套筒
cast	铸
cast iron	铸铁
casting	铸件；铸造
catalog	目录；产品样本
categorize（*v.*）	分类
category（*n.*）	种类，范畴
cathode	阴极
cavity	洞，腔
CCW 或 reverse	反转
cell	电池；（生产）单元
cemented carbide	硬质合金
center	顶尖；打中心孔
center point	圆心
center drill[同]pilot drill	中心钻
centerless grinding	无心磨削
ceramic	陶瓷，陶瓷制品
certify	认证，证明
certificate	证书
chain	链条
chain-type magazine	链式刀库
chamfer	斜面，倒角
channel	通道
character	字符
characteristic	特性
chassis	底盘，机壳
chatter	震颤
check	检查
checkout	检验，校验
chip	芯片；切屑
chip abrasion	切屑粘结
chip breaking	切屑崩碎
chip congestion	切屑聚集
chip conveying	切屑传送
chip conveyor	自动排屑装置
chip disposal	切屑处理
chip ejection	切屑喷出
chip flow	切屑流动
chip removal	切屑排除
chip auger	螺旋排屑器
chuck	卡盘
chuck key	卡盘扳手
chuck jaw	卡盘卡爪
circle	圆
circular interpolation	圆弧插补
circuit	电路
circuit board	电路板
circuit breaker	断路器
circuitry	电路，线路
circumference	圆周，周长
clamp[反] unclamp	夹紧，钳制；夹具，压板
clear chips	清除切屑
clearance	误差，间隙
clearance angle	后角
clockwise	顺时针，正转
closed-loop	闭环

clutch 离合器
cobalt（Co） 钴
code 代码
coded instruction 编码指令
coefficient of friction 摩擦系数
cold-drawn 冷拉
collide（*v.*） 碰撞
collision（*n.*） 碰撞
column 列；立柱
combine（*v.*） 组合
combination（*n.*） 组合
command 命令，指令
command pulse 指令脉冲
comment 注释
commissioning 调试
communication 通信
commutator 换向器，整流器
compact 紧凑的
comparison 比较
compatible with 与……兼容
compatibility 兼容性
compensate（*v.*） 补偿
compensation（*n.*） 补偿
compensating chuck 补偿夹头
compensation table 补偿表
competitive 具有竞争性的
compile cycle 编译循环
complex 复杂的
complexity 复杂性
complicated 复杂的
component 零件，部件
compound rest 复式刀架
compound slide 小滑板
compressor 压缩机
computer 计算机
computerize 计算机化
concentricity 同心度
concept 概念
concurrent 共存的，同时发生的
condensation 浓缩
condenser[近]capacitor 电容器
cone 圆锥体
configure（*v.*） 结构，配置
configuration（*n.*） 结构，配置
configurable 结构的，可配置的
confirm 确认
connect 连接
connector 连接器，插头
console 控制台
constant 常数；恒定的
cutting speed 切削速度
Constant Surface Speed (CSS) 恒表面切削速度
construction 结构，构造
contact 接触
contactor 接触器
contain 包含
continuous-path mode 连续路径方式
contour 轮廓
contour programming 轮廓编程
contouring 轮廓加工
contour milling 轮廓铣
contract 合同
control 控制，控制装置
control panel 控制面板
controller 控制器，控制装置
convenient 方便的
convenience（*n.*） 便利，方便
conventional 常规的
conventional machine tool 普通机床
conversation 对话
conversion 变换，转换
convert 转换
converter 转换器
convey 传送
coolant 冷却液
coolant tank 冷却液箱
coordinate 坐标
coordinate system 坐标系
coordinate measuring machine 坐标测量机
core drill 空芯钻

corona	电晕
corresponding	相应的
cost	成本
cost-effective	节省成本的
counter	计数器
counterbalance	平衡
counterbore	锪孔，镗阶梯孔
counter clockwise	逆时针，反转
countersink	打埋头孔
countersink screw	沉头螺钉
coupled-motion axes	耦合运动轴
coupling	耦合；联轴节
cover	涉及；防护罩
crank	摇手柄
create	创建，建立，生成
criterion[复]criteria	标准
cross	十字；横过
cross feed	横向进给
cross motion	横向运动
cross section	横截面
cross slide	横向滑板
Cubic Boron Nitride (CBN)	立方氮化硼
current	电流
cursor	光标
curvature	弯曲，曲线
cushion time	缓冲时间
customer	客户
customer-friendly	用（客）户友好的
customize	定制
cutter	刀具
cutting tool	切削刀具
cutting edge	切削刃
cutting feed	切削进给
cutting fluid	切削液
cutting force	切削力
cutting oil	切削油
cyber	计算机（网络）的
cycle start	循环启动
cyclical	循环的
cylinder	汽缸，油缸；圆柱体
cylindrical turning	外圆车削

D

damage	损坏
data	数据
data backup	数据备份
data block	数据块
data exchange	数据交换
data processing	数据处理
data storage	数据存储
data transmission	数据传输
database	数据库
deburr	去除毛刺
DC link	直流母线
deactivate	撤销
deceleration	减速
decimal	十进制
decimal point	小数点
decision	决定
decode	译码
decoupling	退耦
decrease[反]increase	减少
dedicated	专门的，专注的
default	默认值
defect	过失，缺点
deflection	偏斜，偏转
define	定义
degree of protection	保护级别
delete	删除
delivery	发送，交付
demonstrate	演示
density	密度
depend on	取决于
depth of cut	切削深度
describe（v.）	描述
description（n.）	描述
deselect	撤销（选择）
design	设计
designate	指定，指派
designation	指定，选派，名称

detect	检测
detector	检测装置
determine	决定，确定
develop	发展
device	设备，装置
devise	设计，发明
diagnosis[复]diagnose	诊断
dial	刻度盘；拨号
dial caliper	带表卡尺
dial indicator	百分表，千分表
diameter	直径
diameter run-out	径向跳动
die	模具
digital	数字的
digital readout	数字显示器
digitization	数字化
digitizer	数字化仪
D/A	数模转换
dimension	尺寸
direction	方向
dirt	脏物
disadvantage	劣势，缺点
disc spring	碟形弹簧
discharge	放电，排出
disk[同]diskette	磁盘，圆盘
disperse	分散，疏散
displacement	位移
display	显示
dissipation	驱散，消耗
distance-to-go	剩余距离，余程
distortion	变形，扭曲
distribution	分配
distribution box	分配盒
distribution switchboard	配电盘
division	分度，刻度
documentation	文件
dog	减速挡块
dovetail	燕尾
dovetailed slide	燕尾式滑板
dovetailed way	燕尾式导轨
download	下载
downtime	停机时间，停工时间
dressing	修整
drift	漂移
drill	钻（孔）；钻头；钻床
drilling machine	钻床
drilling center	钻削中心
drill chuck	钻夹
drill press	台式钻床
drive	驱动
dry run	空运行
dull	钝的
dwell	暂停，停顿
dynamic response	动态响应
dynamo	发电机

E

economical	经济的
edge finder	寻边器
edit	编辑
edit system	编辑系统
effective	有效的
efficient	有效率的
electrical	电的
electrical cabinet	电柜
electrical signal	电信号
electrical pulse	电脉冲
electrician	电工
electrode	电极；电焊条
electromagnetic compatibility	电磁兼容性
electromotive	电动的
electronic device	电子设备
electronic gear	电子齿轮
electronic handwheel	电子手轮
element	要素，元件
ellipse	椭圆
emergency stop	紧急停止
emitter	发射极
employ	雇佣；使用
enable	能使，使能

encode	编码
encoder	编码器
end mill	端铣刀；立铣刀
end point	端点
end product	最终产品
end user	最终用户
end of block	程序结束
energy	能量，能源
engage[反]disengage	啮合
engine lathe	普通车床
engineer	工程师
engrave	雕刻，刻字
engraved scale	刻度尺
enlarge hole	扩孔
ensure	保证
enter	输入，进入
entering angle	切入角，主偏角
entire	整个
environment	环境
equip	装备，配备
equipment	设备
erase	擦除
error	误差；错误
establish	建立
Ethernet	以太网
evaluation	估价
even	偶数；平稳的
ever-rising	持续上升的
exact stop	准确停止
exchangeable	可交换的
execute	执行
expandable	可扩大的
expense	花费
experience	经验
experiment	实验
expertise	专门技术
extend	扩展
extensive	广大的，扩展的
external	外部的

F

fabrication	建造；构造
face mill	面铣刀
facilitate	使容易
facility	设备，设施
facing	车端面
factor	因素，因子
factory	工厂
fall into	分类
fault	故障
feature	特点，特色，功能
feed	进给
feed drive	进给驱动
feed forward	前馈
feedback	反馈
feed function	进给功能，F 功能
feed hold	进给保持
feed rate	进给速度，进给率
feed rod	进给光杠
filament	灯丝
fillet	倒角
file	锉；文件；存档
finish	表面光洁度；精加工
finish boring	精镗
finished part	成品件
finished product	成品
finished workpiece	成品件
finishing cut	精切削
five face machining center	五面加工中心
simultaneous five-axis CNC machine	五轴联动数控机床
fixed stop	固定停止
fixed cycle	固定循环
fixture	夹具
flange	法兰
flash EPROM	快速 EPROM
flexible manufacturing system(FMS)	柔性制造系统
flexible production line	柔性生产线
floppy disk	软盘
flute	凹槽，刀具上的排屑槽

foil	箔
following error	跟随误差
follow-up	跟随
forge	锻造
forging	锻件
form	形状，形式；形成
forming	成形（加工）
forward[近]CW	正转
friction	摩擦
friction clutch	摩擦离合器
fraction	小部分，分数
freely assignable customer key	自定义用户键
frequency	频率
full CNC keyboard	CNC 全键盘
full scale	全刻度
full wave rectified voltage	全波整流电压
full-load [反]no-load	满载
function	功能；起作用
function block	功能块
function call	功能调用
function key	功能键
functionality	功能性
fuse	熔断器
future-oriented	面向未来的
fuzzy control	模糊控制

G

ga(u)ge	计量；计量器
galvanometer	检流计
gantry	龙门
gantry axis	龙门轴
gantry machining center	龙门加工中心
gate	门
gear	齿轮
gear box	齿轮箱
gear change	齿轮切换
gear coupling	齿轮啮合
gear cutting	齿轮切削
gear stage	齿轮级
gear train	齿轮系
general	总的，一般的
general purpose	通用的
general trend	总体趋势
generation	产生
generator	发电机
geometry	几何
geometric shape	几何形状
geometric and positional tolerance	形位公差
graph	图形
graphic programming	图形编程
grating	栅格
grease	油脂
grid	栅格
grind	磨削
grinding machine	磨床
grinding wheel	砂轮
grip	抓紧，握紧
groove	槽
groove sawing	锯槽
grounding	接地
guarantee	保证

H

half-nut	开合螺母
handheld	手持式的
handheld unit	手持单元
handle	把手，手柄
handwheel[近]handle, MPG	手轮
hard disk	硬盘
harden	淬火
hardness	硬度
hardware	硬件
harsh	粗糙的，苛刻的
headstock	床头箱
heat dissipation	散热
heavy-duty machining	强力切削
helical interpolation	螺旋线插补
hexagonal	六角形的
hex-head cap screw	六角螺钉
highlight	加亮
high quality	高质量，优质
high precision	高精度

high speed 高速
High Speed Steel （HSS） 高速钢
high gear range 高速挡
holder and adapter 刀柄和变径套
hole 孔
hole enlarging 扩孔
hollow 中空
hollow-shaft 中空轴
home position 原始位置；参考点位置
horizontal 卧式
horizontal axis 垂直轴
horizontal boring machine 卧式镗床
horizontal machining center(HMC) 卧式加工中心
hose 软管
host computer[近]mainframe 主机
housing 机架，机座
humidity 湿度
hydraulic 液压

I

identical 同样的
identification 证明，同一
identifier 标识符
idle time 空闲时间
image 图像，映像，图片
implement 贯彻
implementation 执行
in a clockwise direction 顺时针方向
in accordance with 和……一致
in addition to 除……之外
in conjunction with 与……协同
in most cases 在多数情况下
in order to 为了
in other words 换句话说
in parallel 并行地，平行地
in relation to 关于，涉及
in short 简而言之
in simple terms 简而言之
in the event of 如果……发生
in the form of 以……形式
in this manner 如此，照这样
inch 英寸
inclined axis 倾斜轴
inclined surface 倾斜表面
incombustible 不能燃烧的
increase 增加
incremental 相对，增量
incremental encoder 增量型编码器
incremental programming 增量编程
index 指数，分度
indexable insert 可转位刀片
indexing head 分度头
indicate 指示
indicator 指示仪表，千分表
individual 个人的，单独的
inertia 惯性，惯量
initialize 初始化
inoperative 无效的，不起作用的
in-process measurement 在线测量
in-process ga(u)ging 在线测量
input 输入
input voltage 输入电压
insert 插入；刀片
inspection 检查，检验
installation 安装
instruction 指令
instruction set 指令集
insulate [近]isolate 绝缘
integrated 综合的，集成的
intelligence 智力，智能
intelligent mode 智能方式
interface 接口
interference 干扰
integrated encoder 内置编码器
interlock 连锁装置；连接，结合
intermediate point 中间点
internal 内部的
Internet 因特网
interpolation 插补；密化，细分，倍频
interrupt 中断

intervention 干涉，干预
inventory 详细目录
inverse 反的
inverter 变频器

J

jam 拥挤，堵塞
jaw 卡爪
jerk 抖动
jig 夹具
job 作业，工作任务
job list 作业表
Jog/JOG 点动，手动
jump 跳变

K

key switch 钥匙开关
keyboard 键盘
keypad 键盘
kinematics 运动学
knee and column 升降台式
know-how 实际知识,专有技术,诀窍
knurl 滚花

L

Ladder 梯形图
lag 滞后
laminate 碾压
large batch production 大批量生产
laser 激光
laser cutting machine 激光切割机
lathe 车床
lathe tool 车刀
layout 布置
lead 导程
lead angle 主偏角
lead terminal 引线接头（轧片）
lead screw 滚珠丝杠
lead screw error 滚珠丝杠误差
least input increment 最小输入增量
length 长度
lengthwise travel 纵向行进
lever 手柄
light duty 轻负荷
lights-out manufacturing 无人值守制造
license 许可证
light source 光源
limit 极限
limit switch 行程开关，限位开关
linear axis 直线轴
linear interpolating 直线插补
linear measurement 直线测量
linear scale 直线光栅尺
linear guide 直线导轨
linear motor 直线电动机
linear way 直线导轨
link 连接；母线
liquid level 液位
live tooling 车铣动力刀座
load 负载；装载
load meter 负载表
local 局部的
local network 局域网
locate 定位，位于
location 位置
logbook 日志
logic 逻辑的
longitudinal 纵向的，长度方向的
lower case letter 小写字母
lube/lubricant 润滑油
lubricate 润滑
lubrication tank 润滑油箱

M

machine 机器，机床；（机）加工
Machine Control Panel（MCP） 机床控制面板
machine control unit 机床控制（数控）装置
machine coordinate system 机床坐标系
machine data 机床数据
machine shop 车间

machine manufacturer	机床制造商	manufacturer	生产厂家，制造商
machine tool	机床	mark	标记，划线
machine tool builder	机床厂	market	市场；销售
machine zero	机床零点	mass production	大批量生产
machinery	机械	master[反]slave	主
machining	机加工	master switch	主（总）开关
machining accuracy	加工精度	match	匹配
machining allowance	加工余量	material	材料，物料
machining center	加工中心	maximum	最大值，上限
machining channel	加工通道	measure	测量
machining plane	加工平面	measuring system	测量系统
machining position	加工位置	mechanical	机械的
machining process	加工过程（工艺）	mechanical engineering industry	机械工程工业
machining practice	加工实践（工艺）	mechanic	机修工
machinist	机械师，机械工	mechanism	机械装置，机构
macro	宏	mechano-electronic	机电的
macro statement	宏程序语句	mech-electronic integration	机电一体化
magazine	刀库	medium [复] media	媒介
magnet valve	电磁阀	meet one's requirements	满足某人的需求
magnetic base	磁性底座	megger	高阻表
magnetic contactor	电磁接触器	memory	存储器
magnetic core	磁芯	memory card	存储卡
magnetic tape	磁带	menu	菜单
main	主要的	merge	合并
main breaker	主断路器；总电源开关	message	信息
main computer	主机	metal	金属
main program	主程序	metal cutting	金属切削
main spindle	主轴	metal removal rates	金属切除率
main spindle motor	主轴电动机	metalworking	金属加工
mainframe	主机	metric	公制的
mains infeed	主电源馈入	micro	微
maintain	保持；维护，保养	microcomputer	微型计算机
maintenance	维护，保养	microelectronics	微电子
management	管理	micrometer	测微计，外径千分尺
manipulator	操作者，机械手	microprocessor	微处理器
manual	手动的；手册	mill	铣；铣刀
manual data input(MDI)	手动数据输入	milling head	铣头
manual programming	手工编程	milling machine	铣床
manually	手动地	minicomputer	小型计算机
manufacture	生产，制造	minimize	最小化

minimum 最小值，下限
minus 负数
mirror 镜像
Miscellaneous function 辅助功能，M 功能
mode 方式
mode key 方式键
modal 模态的
modification 修改
modularity 模块性
modulate 调制
module 模块，组件
mo(u)ld 模具
mold cavity 模腔
mold core 模芯
mold-making 模具制造
monitor 监控
monitoring function 监控功能
monochrome 单色
morse taper shank 莫氏锥柄
motion 运动
motion control unit 运动控制单元
motor 电机
mount[近]install 安装
movement[近]motion 运动
multiply 乘，增加
multi-axis interpolation 多轴插补
multi-dimensional 多维的
multi-edged turning 多刃车削
multimedia 多媒体
multimeter 万用表
multi-point interface 多点接口
multi-tasking machine 多任务机床

N

natural air cooling 自然空气冷却
NC machine tool 数控机床
NC system 数控系统
NC unit 或 MCU 数控装置
negative[反]positive 负的，阴性的
network 网络
neural 中枢的
nipper 钳子
noise-proof 防噪声
no-load 空载
non-chip-producing time 非切削时间
non-circular 非圆的
non-interpolating feed 非插补进给
numeric 数字的
numerical control 数控
numeric-controlled 数控的
nut 螺母

O

odd 奇数
off-center 偏心
offline 离线，机外
offset 偏置
offshore 海外，国外
oil filter 滤油器
on line 在线
online service 在线服务（维修）
open-loop 开环
operate（*v.*） 操作，运行
operation（*n.*） 操作，运行
operator 操作员
operator panel 操作面板
opportunity 机会
optimize 使最优化
optimum 最优的
optimally 最优化地
option 选择，选件
optional stop 选择停止
order 次序；订货
orient（*v.*） 定向，准停
orientation（*n.*） 定向，准停
oriented spindle stop 主轴定向
origin 原点
oscillation 抖动，摆动
output 输出
overcurrent 过流

overhang	伸出部分
overload	过载
override	倍率
override switch	倍率开关
overtravel	超程
overview	概览，概述

P

pack	包装
package	软件包
packing machine	包装机
pallet	托盘
parabola	抛物线
parallel	平行的
parallelism	平行度
parameter	参数
parameterize	参数化
part	零件，工件；切断
part program	加工/零件程序
part surface	工件表面
part zero	工件零点
partial	部分的，局部的
partner	合作伙伴
password	口令
path	路径；轨迹
pattern	曲线，模式
payroll	工资单，计算报告表
PC card	PC 卡
peck drilling	深孔钻
perform	完成
performance	性能
peripheral	外围的；外设
permanent-field	永磁
permissible	允许的
permit	允许
perpendicular	垂直的
perpendicularity	垂直度
personal computer	个人计算机
PTFE guide way	贴塑导轨
photo-electric reader	光电阅读机
photo detector	光敏元件
pinion	小齿轮
pitch	螺距
plain text display	纯文本显示
plane	平面
planning	规划，计划
plasma	等离子
plate part	盘类零件
play-back	录返
plot	打印；图纸
plug	插头；塞子
plug-in	插入式
pneumatic	气动的
pocket milling	槽铣削
polarize	极化
polar coordinate	极坐标
polycrystalline diamond	聚晶金刚石
polynomial interpolation	多项式插补
portable	手提式的，便携式的
position	位置；定位
position coder	位置编码器
position encoder	位置编码器
positioning accuracy	定位精度
positioning axis	定位轴
positioning control	位置控制
positioning error	定位误差
postprocessor	后处理器
potential	电势，电位
potentiometer	电位计
power consumption	功率消耗
power failure	电源故障，断电
power feed	自动进给
power supply	电源
power unit	功率单元
precise	精密的
precision	精度；精密性
precision grinding machine	精密磨床
precision lead screw	精密丝杠
predefined	预定义的
preparatory function	准备功能，G 功能

preprocessing	预处理
prerequisite	先决条件
pre-sales service[反] after-sales service	售前服务
preset	预置
press	压，按
pressworking	压力加工
prestore	预先存储
price	价格；定价
prime	主要的，根本的
primary	主要的；初级的
principle	原则，原理
printed circuit board（PCB）	印制电路板
printer	打印机
printing machine	印刷机
printout	打印输出
printout sheet	打印输出表
priority	先，优先
probe	探针
problem solving	问题解决
processing time	处理时间
produce	生产
product	产品
production	生产
production rate	生产率
production schedule	生产计划
productive	生产的
productivity	生产率
PROFIBUS	一种现场总线
profile	轮廓；仿形
program	编程；程序
program block	程序段
program converter	程序转换器
program preprocessing	程序预处理
program protect	程序保护
program restart	程序再启动
programmable	可编程的
programming device	编程设备
project	计划；发射
property	属性，特性
protection	保护
protection zone	保护区
protocol	协议
prototype	原型
provide	提供
proximity switch	接近开关
pulse encoder	脉冲编码器
pulse train	脉冲序列
punch	冲孔
punched card	冲孔卡
punched tape	冲孔带
push button	按钮

Q

quadrant	象限
quill	钻轴，活动套筒
quality	质量

R

rack	齿条
radius	半径
radial drill	摇臂钻床
rake	（刀具的）前倾面
rake face	前刀面
rake angle	前角
random	随机
random access memory(ROM)	随机存取存储器
rapid	快速的
rapid feed	快速进给
rapid feed override	快速进给修调
rapid traverse	快速进给
rated speed	额定速度
rating	额定值
rating plate	铭牌
ratio	比，比率；系数
raw material	原材料
reaction	反应
reactor	电抗器，扼流圈
read only memory(ROM)	只读存储器
real-time	实时的
ream	铰（孔）

reamer	铰刀
reasonably	适度地
receive	接收
reciprocate	使往复运动
reciprocating axis	往复运动轴
recognize	认出，识别
recompilation	重新编译
record playback	录返
rectangular	矩形的
rectifier	整流器
recurrent	再发生的，循环的
reduce	减少
refer to	参考
reference mark	参考标记
reference point	参考点
reference point approach	回参考点
refinement	精制，精加工
regarding（prep.）	有关
regenerative	再生的
register	登记；寄存器
registered trademark	注册商标
relay	继电器
release	释放
reliable	可靠的
reliability	可靠性
remote diagnosis	远程诊断
remove	转移，取走
repeatability	重复性，再现性；重复定位精度
replace	更换
request	请求
require	需要
requirement	要求
reset	复位
residual material machining	残余材料加工
resolution	分辨率
response	响应
restart	再启动
result	结果
retract（*v.*）	退回，退刀
retraction（*n.*）	收回，退回
retracting strategy	退刀方式，退刀路线
retrieve	重新得到
retrofit	改装
retrofitting（*g.*）	改装
reverse	反转；反向的
revise（*v.*）	修订，修改
revision（*n.*）	修订，修改
revolve（*v.*）	旋转
revolution（*n.*）	旋转；转数
rib	加强筋
ribbon cable	带状电缆
right angle	直角
rigid	刚性的
rigid tapping	刚性攻丝
rigidity [近]stiffness	刚性
robot	机器人
rod	棒
rotary	回转的，旋转的
rotary axis	回转轴
rotary encoder	旋转编码器
rotary table	转台
rotate[近]turn（*v.*）	旋转
rotation（*n.*）	旋转
rotational speed	转速
rotor	转子
rough	粗糙的；粗加工
rough cut	粗加工
roughing reamer	粗加工铰刀
rough turning	粗车
round	圆的；弄圆；四舍五入
round part	圆形零件
out of round	不很圆，失圆
routine	常规，程序
runout	径向跳动
run time	运行时间

S

saddle	床鞍，滑鞍
safety	安全
safety glasses	安全眼镜

safety precaution	安全预防
saw	锯子
scale[近]proportion	比例
scallop	扇形凹口
scallop height	残留高度
scan	扫描
scheduling	计划
scientific research	科学研究
scrap	废料；零碎的
screen	屏幕
screen filter	屏幕滤波器
screen saver	屏幕保护程序
scroll bar	滚动条
sculptured surface	带刻纹的表面
search	检索
secondary	次级的
section	部分
secure	安全的；固定
select	选择
self-hold	自锁
semi-automatic	半自动的
send	发送
sensor	传感器
sequence	顺序
serial	串行
serial interface	串行口
series	系列
service	服务，维修
servo amplifier	伺服放大器
servo control unit	伺服装置
servo enable	伺服能使
servo motor	伺服电动机
servo-drive	伺服驱动
servomechanism	伺服机构
servo system	伺服系统
setpoint	设定点
setting	设置
setting data	设定数据
setup	装夹
shaft	轴
shaft part	轴类零件
shank	柄
shape	形状
shatter	撞碎，毁坏
sheet	（一）张，薄片
sheet metal	金属片
shell part	壳体类零件
shielding	屏蔽
shift	轮班；移动
shock	震动
shop	车间，加工店
shop floor	车间，工场
short-term	短期
shoulder turning	车台阶
signal	信号；发信号
significant	有意义的；重大的
similar to	与……类似
simplified Chinese	简体中文
simplify	简化
simulation	模拟，仿真
simultaneous	同时的，同时发生的
simultaneously	同时地
single block	单程序段
single-axis	单轴
single-spindle	单主轴
sinusoidal	正弦曲线
size[近]dimension	尺寸
skip	跳跃
slant bed	斜床身
slate	石板
slave	副的，从属的
sleeve	套筒，套管，轴套
slide	滑台，滑块
slot	（插）槽，缝
slot milling	铣槽
sneak current	寄生电流，潜行电流
socket	插座，孔
softkey	软键
software	软件
software limit switch	软件限位开关

software package	软件包	standard	标准
Solid State Relay	固态继电器	standardize	使标准化
solenoid valve	电磁阀	start point	起始点
solution	解决办法，方案	start-up	启动，试车，调试
space	空间，太空	state-of-the-art	现代化的，先进的
space age material	太空材料	statement	语句
spare part	备件	statement list	语句表
spare tool	备用刀具	static	静态的
spark	电火花	station	站
special-purpose[反] general purpose	专用的	stator [反] rotor	定子
		status	状态
specific	特定的，具体的	steel plate fabrication	钣金生产
specification	规格，说明	step motor	步进电动机
spectrum	光谱，频谱	stick-slip	爬行
speed[近]velocity	转速；速度	stock	毛坯；存货
speed ratio	速度比率	stock removal	毛坯切除
speed-controlled	速度控制的	storage（*n.*）	存储
speed-dependent	与速度有关的	store（*v.*）	存储
spindle	主轴	straight line	直线
spindle alignment test bar	主轴测试芯棒	straight shank	直柄
spindle function	主轴功能，S 功能	streamline	流线型的
spindle nose	主轴端面	strength	力量
spindle orientation	主轴准停	string	线，一串
spindle override	主轴速度修调	strip	条，片
spindle speed limitation	主轴速度限制	stroke	冲程，行程
spindle range	主轴挡位	structure	结构
spindle taper	主轴锥孔	stub drill	粗短钻
spigot	水龙头，套管	stud	拉钉；双端螺栓
spin	旋转	stylus	测针，触头
spiral	螺旋形	subprogram[近] subroutine	子程序
spline	样条	subsequent	后来的
spline interpolation	样条插补	sub-spindle	副主轴
spot drill	钻中心孔	subtracting	减法
spuareness	方形；垂直	suitable for	适合……的
square wave	方波	supervisory	管理的，监督的
square way	方导轨	supplementary	补充的
squirrel-cage	鼠笼式的	supply	供应，供给
stall	使停转，失速	support	支持
stamp	邮戳，压印	surface finish	表面光洁度
stand for	表示，代表	surface quality	表面质量

surface speed	表面（线）速度
swarf	切屑
switch	开关
switching cabinet	开关柜，强电柜
switchover	切换
swivel	旋转
synchronize（*v.*）	同步
synchronization（*n.*）	同步
synchronized axes	同步轴
synchronous	同步的
synchronous motor	同步电动机
system	系统
system software	系统软件
system variable	系统变量

T

table	工作台；表格
table vise	台虎钳
tachometer	转速表
tailstock	尾架
tailor	定制
tang end	（锉刀）柄脚
tanged shank	扁尾柄
tangential	切线的
tap	丝锥
tapping	攻丝
tape	纸带
tape reader	纸带阅读机
taper	锥度；车锥面
taper shank	锥柄
tapered hole	锥孔
task	任务
teach in(record/play back)	示教
technical	技术的
technical data	技术参数
technique	技术，方法
technology	技术
telescoping steel cover for guideway	导轨防护罩（伸缩式）
temperature	温度
tension	压力，张力
term	名词，术语；期限
terminal	终端
terminal block	接线端子排
terminal screw	端子螺钉
test run	试运行
text editor	文本编辑器
text entry	文本输入
thermal expansion	热膨胀
thermal stability	热稳定性
thread	螺纹
thread cutter	螺纹切削刀具
thread cutting	螺纹切削
thread run-in and run-out	螺纹插入部分与尾部
three-axis movement	三轴运动
three-axis simultaneous movement	三轴联动
three-block chuck	三爪卡盘
through the spindle coolant (TSC)	主轴中心孔冷却
thrust bearing	止推轴承
thrust force	推力，侧向压力
time-out	超时
timer	定时器
time consuming	费时的
toggle	乒乓开关；触发器
toggle key	切换键
tolerance	公差，允差
tool	刀具，工具
tool bit	刀具，刀头
tool blade	刀刃
tool blank	刀坯
tool changing, tool change	换刀
tool function	刀具功能，T 功能
tool holder retention stud	刀柄拉钉
tool identification	刀具识别
tool length offset	刀具长度偏置
tool list	刀具表
tool location	刀位
tool management	刀具管理
tool magazine	刀库
tool offset	刀具偏置

tool path strategy	走刀路线
tool pocket	刀套
tool post	（方）刀架
tool radius compensation	刀具半径补偿
tool release	松刀
tool tip	刀尖
tool type	刀具类型
toolbox	工具盒
toolholder	刀柄
tooling cost	刀具成本
tool-storage magazine	存刀刀库
tool release	松刀
tooling	（用刀具）加工
torque	扭矩，转矩
tough	强硬的，难加工的
track	码道
training	培训
training center	培训中心
training course	培训课程
transfer	传送
transfer line	自动线
transformation	变换，转换
transformer	变压器
transition	转换
translate	翻译，转化
transmission	传动
transmission rate	传输速率
transmit	传输，传送
transparent	透明的，显然的
transverse	横向的
travel	行进；行程
traveling-column machine	动柱式机床
traverse	移动，行进
traverse rate	进给速度
traversing speed	移动速度
trigonometric	三角的
trip	跳闸
troubleshooting	故障寻迹
tunnel	隧道，地道
turning	回转，车
turning center	车削中心
turning machine	车床
turning tool	车刀
turret	刀架
turret head	转塔刀架头
twist drill	麻花钻
twisted cable	绞合电缆
typewrite	打字

U

ultimately	最终地
unclamp	松开
underscore	强调
undershoot[反]overshoot	欠程
unfeasible	不可行的
unidirectional	单向
unit	单元，装置
universal	万能的，通用的
universal interpolator	通用插补器
universal milling head	万能铣头
unload	卸货
upload	上载
upright drill	立式钻床
up to	达到
update	更新
upgrading	上升，升级
up-to-date	最新的
user friendly	用户友好的
user memory	用户存储器
user program	用户程序
user variable	用户变量
utilize	利用

V

valve	阀
variable	变量
variant	类型
various	各种各样的
vector	矢量
velocity	速度

ventilation	通风
Vernier caliper	游标卡尺
versatile	通用的，万能的
version	版本
vertical	立式
vertical axis	竖直轴
vertical boring machine	立式镗床
vertical machining center (VMC)	立式加工中心
vibration	振动
view	观看；视图
vise	虎钳
visual	视觉的，形象的
visualization	形象化，使看得见
voltage	电压
volume	卷，册；体积

W

warranty	保修期
washer	垫圈，衬垫
watchdog	监控器
water cooled	水冷式的
way cover	导轨防护罩
way lube	导轨润滑油
wear	磨损
wearing part	磨损件
website	网站
welding	焊接
width	宽度
winding	绕组，线圈
wiring	接线
wiring duct	走线槽
woodwork	木工活
word	字，字符
workpiece [同]work	工件
workpiece coordinate system	工件坐标系
work schedule	工作计划表
work coordinate offset	工件坐标偏置
work setup	工件装夹
work zero	工件零点
workhold	工件夹持
working area	工作区域
working range	工作范围
workshop	车间，工厂
worldwide	世界范围内的
worn tool	磨损刀具

Z

zero offset	零点偏置
zero speed	零速度
zero return	回零

Appendix D　Technical abbreviations

AC	Alternating Current	交流
AC	Adaptive Control	适应控制
AGV	Automatic Guided Vehicle	自动送料车
APC	Automatic Pallet Changer	自动托盘交换装置
APL	Automatic Parts Loader	自动工件装卸装置
APT	Automatically Programmed Tool	自动编程系统
ASCII	American Standard Code for Information Interchange	美国标准信息交换代码
ATC	Automatic Tool Changer	自动换刀装置
AWF	Automatic Wire Feed	自动走丝
AWG	American Wire Gauge	美国线规
BCD	Binary-Coded Decimal	二进制编码的十进制
CAD	Computer Aided Design	计算机辅助设计
CAM	Computer Aided Manufacturing	计算机辅助制造
CAP	Computer Aided Programming	计算机辅助编程
CBN	Cubic Boron Nitride	立方氮化硼
CCU	Compact Control Unit	紧凑控制单元
CIMS	Computer-Integrated Manufacturing System	计算机集成制造系统
CMOS	Complimentary Metal Oxide Semiconductor	互补金属氧化物半导体
CNC	Computerized Numerical Control	计算机数字控制
CPU	Central Processing Unit	中央处理单元
CRT	Cathode Ray Tube	阴极射线管
CSA	Canadian Standard Association	加拿大标准协会
CSS	Constant Surface Speed	恒表面速度
CW	Clockwise	顺时针旋转，正转
CCW	Counter Clockwise	逆时针旋转，反转
DC	Direct Current	直流
DCS	Distributed Control System	集散系统
DGNOS	Diagnostic	诊断
DIR	Directory	目录
DNC	Direct Numerical Control/Distributed Numerical Control	直接数字控制/分布式数字控制
DOS	Disk Operating System	磁盘操作系统
DRF	Differential Resolver Function	微分解算功能
DRO	Digital ReadOut	数显装置

EDM	Electricity Discharge Machine	电火花加工机
EIA	Electronics Industries Association	美国电子工业协会
EOB	End of Block	程序段结束
EOF	End of File	文件结束
E-Stop	Emergency Stop	紧急停止
EPROM	Erasable Programmable Read Only Memory	可擦除可编程只读存储器
FA	Factory Automation	工厂自动化
FEPROM	Flash Erasable Programmable Read Only Memory	快闪 EPROM
FIFO	First In First Out	先进先出
FMC	Flexible Manufacturing Cell	柔性制造单元
FMS	Flexible Manufacturing System	柔性制造系统
FWD	Forward	向前，正转
GT	Group Technology	成组工艺
HMC	Horizontal Machining Center	卧式加工中心
HP	HorsePower	马力，功率
HSS	High Speed Steel	高速钢
IC	Integrated Circuit	集成电路
ID	Inside Diameter	内径
IEEE	Institute of Electrical and Electronic Engineers	美国电气电子工程师协会
IGBT	Isolated Gate Bipolar Transistor	绝缘门双极晶体管
IN	Inch	英寸
ISO	International Standard Organization	国际标准化组织
JIT	Just-In-Time	准时生产制
LAN	Local Area Network	局域网络
LED	Light Emitting Diode	发光二极管
LF	Line Feed	跳至下一行
LSIC	Large Scale Integrated Circuit	大规模集成电路
MAP	Manufacturing Automation Protocol	制造自动化协议
MCS	Machine Coordinate System	机床坐标系
MCU	Machine Control Unit	机床控制单元
MDA	Manual Data Automatic	手动数据自动
MDI	Manual Data Input	手动数据输入
MIT	the Massachusetts Institute of Technology	麻省理工学院
MMC	Man Machine Communication	人机通信
MODEM	Modulation and DEModulation	调制解调器
MPG	Manual Pulse Generator	手摇脉冲发生器

MPP	Manufacturing Process Planning	制造工艺规划
MRP	Manufacturing Resource Planning	生产资源规划
MTB	Machine Tool Builder	机床制造商
NC	Numerical Control	数控
NC	Normally Closed	常闭（触点）
NO	Normally Open	常开（触点）
NURBS	Non-Uniform Rational B Splines	非一致有理化 B 样条
OD	Outside Diameter	外径
OEM	Original Equipment Manufacturer	原始设备制造商
OPER	Operator	操作员
PARAM	Parameter	参数
PC	Personal Computer	个人计算机
PCB	Printed Circuit Board	印制电路板
PCD	PolyCrystalline Diamond	聚晶金刚石
PCMCIA	Personal Computer Memory Card International Association	国际个人计算机存储卡协会
PCS	Process Control System	过程控制系统
PLC	Programmable Logical Controller	可编程逻辑控制器
POSIT	Position	位置
PROG	Program	程序
PWM	Pulse Width Modulation	脉冲宽度调制
RAM	Random Access Memory	随机存取存储器
RET	Return	返回
ROM	Read Only Memory	只读存储器
RPM	Revolution Per Minute	转/分
RS232	Recommended Standard	EIA 颁布的一种通信标准
SFM	Surface Feet per Minute	线速度，英尺/分
STL	Statement List	语句表
SV	Servo	伺服
TCP/IP	Transmission Control Protocol/Interconnect Protocol	传输控制协议/互连协议
TOP	Technical Office Protocol	办公室自动化协议
VMC	Vertical Machining Center	立式加工中心
WAN	Wide Area Network	广域网
WCS	Work Coordinate System	工件坐标系
WYSIWYG	What You See Is What You Get	所见即所得

Grammatical abbreviations

n.	noun	名词
v.	verb	动词
a.	adjective	形容词
ad.	adverb	副词
g.	gerund	动名词
pl.	plural	复数
anto.	antonym	反义词
syno.	synonym	同义词
fig.	figure	图片
ref.	reference	参考

Appendix E List of G Codes and M Codes

G code	Group	Meaning	
G00	01*	Rapid Motion	（快速定位）
G01	01	Linear Interpolation Motion	（直线插补运动）
G02	01	CW Interpolation Motion	（顺圆插补运动）
G03	01	CCW Interpolation Motion	（逆圆插补运动）
G04	00	Dwell	（暂停）
G09	00	Exact Stop	（准确停止）
G10	00	Data setting	（数据设定）
G11	00	Data setting cancel	（取消数据设定）
G12	00	CW Circular Pocket Milling (Yasnac)	（顺圆槽铣削）
G13	00	CCW Circular Pocket Milling (Yasnac)	（逆圆槽铣削）
G17	02*	XY Plane Selection	（XY平面选择）
G18	02	ZX Plane Selection	（ZX平面选择）
G19	02	YZ Plane Selection	（YZ平面选择）
G20	06*	Select Inches	（选择英制）
G21	06	Select Metric	（选择公制）
G28	00	Return To Reference Point	（回到参考点）
G29	00	Return From Reference Point	（从参考点返回）
G31	00	Feed Until Skip (optional)	（跳转）
G35	00	Automatic Tool Diameter Measurement	（自动刀具直径测量）
G36	00	Automatic Work Offset Measurement	（自动工件零点偏置测量）
G37	00	Automatic Tool Offset Measurement	（自动刀具偏置测量）
G40	07*	Cutter Compensation Cancel	（取消刀具补偿）
G41	07	Cutter Compensation Left	（刀具左补偿）
G42	07	Cutter Compensation Right	（刀具右补偿）
G43	08	Tool Length Compensation +	（刀具长度补偿+）
G44	08	Tool Length Compensation -	（刀具长度补偿-）
G47	00	Text Engraving	（刻字）
G49	08*	G43/G44 Cancel	（取消G43/G44）
G50	11	G51 Cancel	（取消G51）
G51	11	Scaling (optional)	（比例缩放）
G52	12	Select Work Coordinate System G52 (Yasnac)	（选择工件坐标系）
G52	00	Set Local Coordinate System (Fanuc)	（设定局部坐标系）
G52	00	Set Local Coordinate System (HAAS)	（设定局部坐标系）
G53	00	Non-Modal Machine Coordinate Selection	（非模态机床坐标系选择）
G54	12*	Select Work Coordinate System 1	（设定工件坐标系1）
G55	12	Select Work Coordinate System 2	（设定工件坐标系2）
G56	12	Select Work Coordinate System 3	（设定工件坐标系3）
G57	12	Select Work Coordinate System 4	（设定工件坐标系4）

G58	12	Select Work Coordinate System 5	（设定工件坐标系5）
G59	12	Select Work Coordinate System 6	（设定工件坐标系6）
G60	00	Unidirectional Positioning	（单一方向定位）
G61	13	Exact Stop Modal	（模态准确停止）
G64	13*	G61 Cancel	（取消G61）
G65	00	Macro Subroutine Call (optional)	（调用宏程序）
G68	16	Rotation (optional)	（旋转）
G69	16	G68 Cancel (optional)	（取消G68）
G70	00	Bolt Hole Circle (Yasnac)	（螺栓孔圆周排列）
G71	00	Bolt Hole Arc (Yasnac)	（螺栓孔圆弧排列）
G72	00	Bolt Holes Along an Angle (Yasnac)	（螺栓孔斜线排列）
G73	09	High Speed Peck Drill Canned Cycle	（高速深孔钻固定循环）
G74	09	Reverse Tap Canned Cycle	（反攻丝固定循环）
G76	09	Fine Boring Canned Cycle	（精镗固定循环）
G77	09	Back Bore Canned Cycle	（背镗固定循环）
G80	09*	Canned Cycle Cancel	（取消固定循环）
G81	09	Drill Canned Cycle	（钻削固定循环）
G82	09	Spot Drill Canned Cycle	（钻孔固定循环，孔底暂停）
G83	09	Normal Peck Drill Canned Cycle	（普通深孔钻固定循环）
G84	09	Tapping Canned Cycle	（攻丝固定循环）
G85	09	Boring Canned Cycle	（镗孔固定循环）
G86	09	Bore/Stop Canned Cycle	（镗孔固定循环，孔底主轴停止）
G87	09	Bore/Stop/Manual Retract Canned Cycle	（反镗固定循环）
G88	09	Bore/Dwell/Manual Retract Canned Cycle	（镗孔固定循环，孔底暂停后，主轴停止）
G89	09	Bore and Dwell Canned Cycle	（镗孔固定循环，孔底暂停，主轴不停）
G90	03*	Absolute	（绝对编程方式）
G91	03	Incremental	（增量编程方式）
G92	00	Set Work Coordinates	（设定工件坐标系）
G94	05*	Feed Per Minute Mode	（每分进给方式）
G98	10*	Initial Point Return	（返回初始点平面）
G99	10	R Plane Return	（返回R点平面）

注：带*的代码为开机默认设置

M code	Meaning	
M00	Stop Program	（程序停止）
M01	Optional Program Stop	（程序选择性停止）
M02	Program End	（程序结束）
M03	Spindle Forward	（主轴正转）
M04	Spindle Reverse	（主轴反转）
M05	Spindle Stop	（主轴停转）
M06	Tool Change	（自动换刀）
M07	Coolant On(Flood)	（冷却液开启（水状））
M08	Coolant On(Mist)	（冷却液开启（雾状））
M09	Coolant Off	（冷却液关闭）
M10	Engage 4th Axis Brake	（第4轴制动器啮合）
M11	Release 4th Axis Brake	（第4轴制动器释放）
M12	Engage 5th Axis Brake	（第5轴制动器啮合）
M13	Release 5th Axis Brake	（第5轴制动器释放）
M16	Tool Change (same as M06)	（自动换刀）
M17	End of Subprogram	（子程序结束）
M19	Orient Spindle	（主轴定向）
M21-M28	Optional Pulsed User M Function with Fin	（用户M功能）
M30	Program End and Rewind	（程序结束并反绕）
M31	Chip Conveyor Forward	（排屑器正转）
M32	Chip Conveyor Reverse	（排屑器反转）
M33	Chip Conveyor Stop	（排屑器停转）
M34	Increment Coolant Spigot Position	（冷却液喷嘴位置上升）
M35	Decrement Coolant Spigot Position	（冷却液喷嘴位置下降）
M36	Pallet Rotate	（托盘回转）
M39	Rotate Tool Turret	（刀具转塔架回转）
M41	Low Gear Override	（切换低档齿轮）
M42	High Gear Override	（切换高档齿轮）
M50	Execute Pallet Change	（执行托盘交换）
M51-M58	Set Optional User M	（设定可选M功能）
M61-M68	Clear Optional User M	（取消可选M功能）
M75	Set G35 or G136 Reference Point	（设定G35或G136参考点）
M76	Disable Displays	（禁止显示）
M77	Enable Displays	（允许显示）
M78	Alarm if skip signal found	（有跳转信号则报警）
M79	Alarm if skip signal not found	（无跳转信号则报警）
M82	Tool Unclamp	（刀具松开）
M86	Tool Clamp	（刀具夹紧）
M88	Through the Spindle Coolant ON	（主轴中心孔冷却开启）

M89	Through the Spindle Coolant OFF	（主轴中心孔冷却关闭）
M95	Sleep Mode	（睡眠模式）
M96	Jump if no Input	（无输入则跳转）
M97	Local Sub-Program Call	（局部子程序调用）
M98	Sub Program Call	（子程序调用）
M99	Sub Program Return Or Loop	（子程序返回或循环执行）

References

[1] HAAS Automation. Operator's manual and service manual. the USA: 2002.

[2] FANUC. Operator's manual. Japan: 2000.

[3] The Army Institute for Professional Development. Lathe operations. U.S: 2007.

[4] www.mini-lathe.com

[5] www.haascnc.com

[6] www.hardinge.com

[7] www.mazak.com

[8] www.ad.siemens.com

[9] www.toolingu.com

[10] 朱一纶. 电子技术专业英语. 北京: 电子工业出版社，2003.

[11] 刘瑛等. 数控技术英语. 北京: 化学工业出版社，2003.

[12] 汤彩萍. 数控技术专业英语（第 3 版）. 北京: 电子工业出版社，2013.